Nanotechnology Research Directions: IWGN Workshop Report

About the National Science and Technology Council

President Clinton established the National Science and Technology Council (NSTC) by Executive Order on November 23, 1993. This cabinet-level council is the principal means for the President to coordinate science, space and technology policies across the Federal Government. NSTC acts as a "virtual" agency for science and technology (S&T) to coordinate the diverse parts of the Federal research and development (R&D) enterprise. The NSTC is chaired by the President. Membership consists of the Vice President, Assistant to the President for Science and Technology, Cabinet secretaries and agency heads with significant S&T responsibilities, and other White House officials.

An important objective of the NSTC is the establishment of clear national goals for Federal S&T investments in areas ranging from information technologies and health research, to improving transportation systems and strengthening fundamental research. The Council prepares R&D strategies that are coordinated across Federal agencies to form an investment package that is aimed at accomplishing multiple national goals.

To obtain additional information regarding the NSTC, contact 202-456-6100 or see the NSTC Web site at
http://www.whitehouse.gov/WH/EOP/OSTP/NSTC/html/NSTC_Home.html.

Interagency Working Group on Nanoscience, Engineering and Technology (IWGN)

Chair: M.C. Roco, NSF
White House IWGN Co-chair: T.A. Kalil, Special Assistant to the President, W.H. Economic Council
Vice-chair: R. Trew, DOD
Executive Secretary: J.S. Murday, NRL
Members:
White House: T.A. Kalil
OSTP: K. Kirkpatrick
NSTC: J. Porter
OMB: D. Radzanowski
DOC: P. Genther-Yoshida, M.P. Casassa (NIST), R.D. Shull (NIST)
DOD: R. Trew, J.S. Murday (NRL), G.S. Pomrenke (AFOSR)
DOE: I.L. Thomas, R. Price, B.G. Volintine
DOT: R.R. John, A. Lacombe (Volpe Center)
DoTREAS: E. Murphy
NASA: S. Venneri, G.H. Mucklow, M. Meyyappan (NASA Ames), T. Krabach (JPL)
NIH: J.A. Schloss, E. Kousvelari
NSF: M.C. Roco, T.A. Weber, M.P. Henkart
Public Affairs Consultant: J. Canton

International Technology Research Institute, World Technology (WTEC) Division, Loyola College

R.D. Shelton, ITRI Director
G.M. Holdridge, WTEC Division Director and ITRI Series Editor

P. Johnson, Editorial Assistant

National Science and Technology Council (NSTC)
Committee on Technology (CT)
Interagency Working Group on Nanoscience, Engineering and Technology
(IWGN)

Nanotechnology Research Directions: IWGN Workshop Report

Vision for Nanotechnology R&D in the Next Decade

On behalf of NSTC/CT/IWGN
Edited by M.C. Roco (IWGN Chair),
R.S. Williams (private sector), and P. Alivisatos (academe)

KLUWER ACADEMIC PUBLISHERS
DORDRECHT / BOSTON / LONDON

Library of Congress Cataloging-in-Publication Data

ISBN 978-90-481-5416-6

Published by Kluwer Academic Publishers,
P.O. Box 17, 3300 AA Dordrecht, The Netherlands.

Sold and distributed in North, Central and South America
by Kluwer Academic Publishers,
101 Philip Drive, Norwell, MA 02061, U.S.A.

In all other countries, sold and distributed
by Kluwer Academic Publishers,
P.O. Box 322, 3300 AH Dordrecht, The Netherlands.

Reprinted 2001

Printed on acid-free paper

This report was prepared under the guidance of NSTC/CT. Any opinions, conclusions or recommendations expressed in this material are those of the Interagency Working Group on Nanoscience, Engineering and Technology, and do not necessarily reflect the views of the Administration or individual funding agencies.

International Technology Research Institute, World Technology (WTEC) Division

This document is available on the Internet at
http://itri.loyola.edu/nano/IWGN.Research.Directions/

Nanotechnology Research Directions: IWGN Workshop Report

Vision for Nanotechnology R&D in the Next Decade

Table of Contents

EXECUTIVE SUMMARY

Nanotechnology is the creation and utilization of materials, devices, and systems through the control of matter on the nanometer-length scale, that is, at the level of atoms, molecules, and supramolecular structures. The essence of nanotechnology is the ability to work at these levels to generate larger structures with fundamentally new molecular organization. These "nanostructures," made with building blocks understood from first principles, are the smallest human-made objects, and they exhibit novel physical, chemical, and biological properties and phenomena. The aim of nanotechnology is to learn to exploit these properties and efficiently manufacture and employ the structures.

Control of matter on the nanoscale already plays an important role in scientific disciplines as diverse as physics, chemistry, materials science, biology, medicine, engineering, and computer simulation. For example, it has been shown that carbon nanotubes are ten times as strong as steel with one sixth of the weight, and that nanoparticles can target and kill cancer cells. Nanoscale systems have the potential to make supersonic transport cost-effective and to increase computer efficiency by millions of times. As understanding develops of the way natural and living systems are governed by molecular behavior at nanometer scale, and as this understanding begins to be felt in science and medicine, researchers seek systematic approaches for nanoscale-based manufacturing of human-made products.

All natural materials and systems establish their foundation at the nanoscale; control of matter at molecular levels means tailoring the fundamental properties, phenomena, and processes exactly at the scale where the basic properties are determined. Therefore, by determining the novel properties of materials and systems at this scale, nanotechnology could impact the production of virtually every human-made object—everything from automobiles, tires, and computer circuits to advanced medicines and tissue replacements—and lead to the invention of objects yet to be imagined. Nanotechnology will be a strategic branch of science and engineering for the next century, one that will fundamentally restructure the technologies currently used for manufacturing, medicine, defense, energy production, environmental management, transportation, communication, computation, and education.

As the twenty-first century unfolds, nanotechnology's impact on the health, wealth, and security of the world's people is expected to be at least as significant as the combined influences in this century of antibiotics, the integrated circuit, and human-made polymers. Dr. Neal Lane, Advisor to the President for Science and Technology and former National Science Foundation (NSF) director, stated at a Congressional hearing in April 1998, "If I were asked for an area of science and engineering that will most likely produce the breakthroughs of tomorrow, I would point to nanoscale science and engineering." Recognizing this potential, the White House Office of Science and Technology Policy (OSTP) and the Office of Management and Budget (OMB) have issued a joint memorandum to Federal agency heads that identifies nanotechnology as a research priority area for Federal investment in fiscal year 2001.

This report charts "Nanotechnology Research Directions," as developed by the Interagency Working Group on Nano Science, Engineering, and Technology (IWGN) of the National Science and Technology Council (NSTC). The report incorporates the views of leading experts from government, academia, and the private sector. It reflects the consensus reached at an IWGN-sponsored workshop held on January 27–29, 1999, and detailed in contributions submitted thereafter by members of the U.S. science and engineering community. (See Appendix A for a list of contributors.) This report describes challenges that are posed and opportunities that are offered by nanotechnology and outlines the steps we must take as a nation if we are to benefit from the advances that are envisioned. Moreover, it proposes a national nanotechnology initiative consistent with the OSTP/OMB memorandum. This emphasizes three crucial areas: developing a balanced research and development infrastructure, advancing critical research areas, and nurturing the scientific and technical workforce of the next century. The initiative proposes doubling the Federal investment in nanotechnology and founding a cooperative grand alliance of government, academia, and the private sector to promote U.S. world leadership in nanotechnology.

SYNOPSIS OF RECOMMENDATIONS

Workshop participants agreed that the benefits of nanotechnology could best be realized through a cooperative national program involving universities, industry, government agencies at all levels, and the government/national laboratories. To address the scientific and technological challenges and reap nanotechnology's social and economic benefits, workshop participants recommended a national initiative with the following objectives:

- Support long-term nanoscience and engineering research leading to fundamental discoveries of novel phenomena, processes, and tools
- Improve institutional structures so they foster and nourish developments
- Encourage the type of transdisciplinary and multi-institutional cooperation required in this new area
- Provide new types of educational opportunities to train the nanotechnologists and entrepreneurs of the future
- Create the physical infrastructure to enable first-class basic research, exploration of applications, development of new industries, and rapid commercialization of innovations

Within their vision of a "grand coalition" contributing to a national nanotechnology initiative, workshop participants proposed specific objectives for academe, private industry, Government laboratories, Government funding agencies, and professional science and engineering societies, as follows:

1. Academe

- Promote interdisciplinary work involving multiple departments
- Foster on-campus nanotechnology centers for greater interaction
- Introduce nanoscience and engineering in existing and new courses
- Create or connect "regional coalitions" that involve industry/ technology generation
- Ease intellectual property restrictions to improve information flow with industry
- Establish graduate and postdoctoral fellowships for interdisciplinary work

2. Private Sector

- Build up investment by maintaining in-house research activities in nanotechnology
- Join, contribute to, or lead regional coalitions for precompetitive nanotechnology research and information dissemination
- Sponsor nanotechnology startups/spin-offs

3. Government R&D Laboratories

- Pursue applications of nanotechnology in support of respective agency missions
- Join regional coalitions with universities and industry, and cultivate information flow
- Provide unique measurement and manufacturing capabilities at nanoscale facilities (synchrotrons, microscopy centers, etc.)
- Provide measurement standards for the nanotechnology field

4. Government Funding Agencies

- Establish a national nanotechnology initiative in fiscal year 2001 that will approximately double the current Government annual investment of about $255 million (in fiscal year 1999) in R&D supporting nanoscience, engineering and technology
- Emphasize small, transdisciplinary research groups in academe within and among universities, and promote policies that foster collaboration between academe, private sector, and government laboratories
- Support nanoscience and engineering fellowships that are not tied to one discipline
- Develop and maintain an information system and databases specifically for nanoscience and engineering available to the community at large to serve rapid development of research and education in the field
- Sponsor regional university and Government lab centers in partnership with industry to cultivate exploratory research, shared research in critical areas, education and information flow
- Establish "vertical centers" where fundamental research, applied research, technology development, and prototype construction or clinical evaluations can be pursued concurrently

- Promote international collaborations for cost-sharing and joint centers/networks of excellence, where appropriate, for fundamental studies

5. Professional Science and Engineering Societies

- Establish interdisciplinary forums that accelerate progress in research and development in nanoscience, engineering and technology, and facilitate its transition into other technologies
- Convene groups of scientists and engineers who have not collaborated traditionally
- Reach out to the international research communities to ensure U.S. awareness of the latest advances
- Develop symposia to explore educational opportunities at K-12, undergraduate, and graduate levels
- Invite industrial players to participate in interdisciplinary job fairs and interview prospective scientists and engineers for nano-related openings

On behalf of the Interagency Working Group on Nanoscience, Engineering and Technology (IWGN),

Dr. M.C. Roco, National Science Foundation, Chair of the IWGN

Dr. R.S. Williams, Hewlett-Packard Co. (representing the private sector)

Dr. P. Alivisatos, University of California, Berkeley (representing academe)

TECHNICAL SUMMARY

The National Science and Technology Council's Interagency Working Group on Nano Science, Engineering, and Technology (IWGN) held a workshop on January 27–29, 1999, to survey research and development as well as education opportunities in nanoscience, engineering and technology, examine what opportunities exist, develop a baseline understanding of the Federal role, and ascertain what is required to ensure that the United States benefits from this new field. Participants at the workshop and other contributors after the meeting represented academic, industrial, and Government organizations and a range of disciplines, including biology, chemistry, materials science, physics, and engineering.

From workshop presentations it was clear that the on-going discovery of novel phenomena and processes at the nanometer scale is providing science with a wide range of tools, materials, devices, and systems with unique characteristics. By using structure at the nanoscale as a physical variable, it is possible to greatly expand the range of performance of existing chemicals and materials. Scientists can already foresee using patterned monolayers for a new generation of chemical and biological sensors; nanoscale switching devices to improve computer storage capacity by a factor of a million; tiny medical probes that will not damage tissues; entirely new drug and gene delivery systems; nanostructured ceramics, polymers, metals, and other materials with greatly improved mechanical properties; nanoparticle-reinforced polymers in lighter cars; and nanostructured silicates and polymers as better contaminant scavengers for a cleaner environment. Current research is moving rapidly from observation and discovery to design and fabrication of complex nanoscale assemblies. Soon, a systems approach grounded in multidisciplinary research will be required for continued and rapid progress.

Workshop participants—all respected experts in the nanotechnology field—emphasized the breadth and variety of applications and the common obstacles facing extremely disparate research areas. They frequently noted nanotechnology's potential to displace major existing technologies, create new industries, and transform archetypal scientific models in the areas of energy, environment, communications, computing, medicine, space exploration, national security, and any area based on materials. However, while recognizing nanotechnology's potential to spawn an industrial revolution in coming decades, the consensus was that the challenges ahead in basic discovery, invention, and eventual manufacturing are formidable. New methods of investigation at the nanoscale, novel scientific theories, and different fabrication paradigms are critical.

The main objectives of the IWGN workshop and this report were to identify science and technology paradigm changes underway as a result of nanoscale research and development; current and potential applications of nanotechnology; and means to strengthen the U.S. research and development infrastructure to capture the potential of nanotechnology in the next decade.

After an introduction for non-specialists, Chapters 1-3 of this report outline the fundamental scientific challenges in nanotechnology and the investigative tools that have made possible the development of this field. Chapter 4 surveys current developments and visionary perspectives for synthesis and assembly of nanostructures; Chapters 5-10 survey the main areas of nanotechnology application; Chapter 11 describes future infrastructure needs for research and development and education as compared to the present; and Chapter 12 analyzes roles, priorities, and strategies for U.S. funding agencies. Recommendations for academe, the private sector, Government R&D laboratories, Government funding agencies, and professional societies begin on page xix.

Definition of Nanotechnology

Nanotechnology is the popular term for the construction and utilization of functional structures with at least one characteristic dimension measured in nanometers. Such materials and systems can be rationally designed to exhibit novel and significantly improved physical, chemical, and biological properties, phenomena, and processes because of their size. When characteristic structural features are intermediate in extent between isolated atoms and bulk materials, in the range of about 10^{-9} to 10^{-7} m (1 to 100 nm), the objects often display physical attributes substantially different from those displayed by either atoms or bulk materials.

Properties of matter at the nanoscale are not necessarily predictable from those observed at larger scales. Important changes in behavior are caused not only by continuous modification of characteristics with diminishing size, but also by the emergence of totally new phenomena such as quantum size confinement, wave-like transport, and predominance of interfacial phenomena. Once it is possible to control feature size and shape, it is also possible to enhance material properties and device functions beyond what are already established. Currently known nanostructures include such remarkable entities as carbon nanotubes, proteins, DNA, and single-electron transistors that operate at room temperature. Rational fabrication and integration of nanoscale materials and devices herald a revolutionary age for science and technology, provided we can discover and fully utilize their underlying principles.

A Revolution at the Limits of the Physically Possible

In 1959 Nobel laureate physicist Richard Feynman delivered his now famous lecture, "There is Plenty of Room at the Bottom."[1] He stimulated his audience with the vision of exciting new discoveries if one could fabricate materials and devices at the atomic/molecular scale. He pointed out that, for this to happen, a new class of miniaturized instrumentation would be needed to manipulate and measure the properties of these small—"nano"—structures.

It was not until the 1980s that instruments were invented with the capabilities Feynman envisioned. These instruments, including scanning tunneling microscopes, atomic force microscopes, and near-field microscopes, provide the "eyes" and "fingers" required for nanostructure measurement and manipulation. In a parallel development, expansion of computational capability now enables sophisticated simulations of material behavior at the nanoscale. These new tools and techniques have sparked excitement throughout the scientific community. Scientists from many disciplines are now avidly fabricating and analyzing nanostructures to discover novel phenomena based on structures with at least one dimension under the "critical scale length" of 100 nm. Nanostructures offer a new

[1] Published later: Feynman, R.P. 1961. There is plenty of room at the bottom. In *Miniaturization*. New York: Reinhold.

paradigm for materials manufacture by submicron-scale assembly (ideally, utilizing self-organization and self-assembly) to create entities from the "bottom up" rather than the "top down" ultraminiaturization method of chiseling smaller structures from larger ones. However, we are just beginning to understand some of the principles to use to create "by design" nanostructures and how to economically fabricate nanodevices and systems. Second, even when fabricated, the physical/chemical properties of those nanostructured devices are just beginning to be uncovered; the present micro- and larger devices are based on models working only at scale lengths over the 100+ nm range. Each significant advance in understanding the physical/chemical properties and fabrication principles, as well as in development of predictive methods to control them, is likely to lead to major advances in our ability to design, fabricate and assemble the nanostructures and nanodevices into a working system.

What the Visionaries Say

John Armstrong, formerly Chief Scientist of IBM, wrote in 1991, "I believe nanoscience and nanotechnology will be central to the next epoch of the information age, and will be as revolutionary as science and technology at the micron scale have been since the early '70s." More recently, industry leaders, including those at the IWGN workshop, have extended this vision by concluding that nanoscience and technology have the potential to change the nature of almost every human-made object in the next century. They expect significant improvements in materials performance and changes in manufacturing to lead to a series of revolutionary changes in industry.

At the workshop, Horst Stormer, Nobel Laureate, articulated the vision many share: "Nanotechnology has given us the tools. . . to play with the ultimate toy box of nature — atoms and molecules. Every thing is made from it. The combination of our top-down tools and methods with self-assembly on the atomic scale provides an impressive array of novel opportunities to mix-and-match hunks of chemistry and biology with artificially defined, person-made structures. The possibilities to create new things appear limitless."

George Whitesides, Professor of Chemistry at Harvard, in 1998 gave information storage as an example of the radical changes nanotechnology could make possible: "You could [with nanodevices] get, in something the size of a wristwatch, the equivalent of 1,000 CDs. That starts approaching a fraction of the reference library that you need for your life...It's one of those ideas that shifts a little bit the notion of how a life should be led."[2]

Although considerable uncertainty is prevalent in predicting future benefits of investments, this report attempts to anticipate benefits that will most likely occur within a few decades. A significant lesson from the 20th century is that predictions of the state of a particular technology several decades in the future often fall far short of what is actually accomplished because one foresees only evolutionary changes, while scientific and technological revolutions are almost impossible to predict (see "Introduction to Nanotechnology for Nonspecialists," page xxv).

Fundamental Science Issues to be Explored

The investigative tools and level of understanding of basic nanoscale phenomena are now only rudimentary. For the promise of nanotechnology to be realized, much more fundamental scientific knowledge is needed, including understanding of molecular self-organization, how to construct quantum devices, and how complex nanostructure systems operate. It is difficult at this moment to make sharp distinctions between fundamental and applied science in nanotechnology. This is not a new phenomenon: recall that the discovery of the laser revolutionized several fields, including both communications and surgery, while the basic scientific principles were still being investigated. There are several areas in physics, chemistry, materials science, electrical engineering, and other disciplines where the basic sciences must be thoroughly developed before a concrete nanotechnology will have the chance to emerge. Several broad, transdisciplinary questions being asked in current fundamental nanoscience R&D illustrate the challenge:

1. What new and novel quantum properties will be enabled by nanostructures, especially at room temperatures?

---

[2] Whitesides, G.M. 1998. Nanotechnology: Art of the possible. *Technology Review* November/ December.

2. How different from bulk behavior will be the properties of interfacial regions between contiguous nanostructures? How can new technologies exploit these properties?
3. What are the surface reconstructions and rearrangements of atoms in nanocrystals and nanorods? Is it possible to prepare epitaxial core-shell systems in nanocrystals?
4. Can carbon nanotubes of a single length and helicity be synthesized and purified as isolated molecular species? Is it possible to reproducibly prepare heterojunctions in one-dimensional nanostructures?
5. What new insights in our understanding of complex polymer, supramolecular, and biological systems will come from the capability to examine single-molecule properties?
6. How extensively can one use parallel self-assembly techniques to control the relative arrangements of nanoscale components according to a complex, designed sequence before error rates become unacceptable?
7. Are there processes that would lead to economic preparation of nanostructures with the control of size, shape, composition, and surface states necessary for advanced device applications?

The Societal and Economic Impacts of Nanotechnology

Potential applications of nanotechnology are pervasive, in the fields described below:

Materials and Manufacturing

Nanotechnology is fundamentally changing the way materials and devices will be produced in the future. The ability to synthesize nanoscale building blocks with precisely controlled size and composition and then to assemble them into larger structures with unique properties and functions will revolutionize segments of the materials manufacturing industry. Nanostructuring is expected to bring about lighter, stronger, and programmable materials; reductions in life-cycle costs through lower failure rates; innovative devices based on new principles and architectures; and use of molecular/cluster manufacturing.

Molecular/cluster manufacturing takes advantage of assembly at the nanoscale level for a given purpose. Structures not previously observed in nature can be developed. Challenges include synthesis of materials by design, development of bio- and bio-inspired materials, development of cost-effective and scalable production techniques, and determination of the nanoscale initiators of materials failure. Applications of nanotechnology to materials and manufacturing include the following:

- Forming nanostructured metals and ceramics at exact shapes without machining
- Improved color printing brought about by nanometer-scale particles that have the best properties of both dyes and pigments
- Nanoscale cemented and plated carbide materials and nanocoatings for cutting tools, and other electronic, chemical, and structural applications
- New standards for measurements at nanoscale
- Nanofabrication on a chip with high levels of complexity and functionality

Nanoelectronics and Computer Technology

The Semiconductor Industry Association (SIA) has developed a roadmap for continued enhancements in miniaturization, speed, and power reduction in information processing devices, e.g. sensors for signal acquisition, processors, memories and displays. The SIA's 1997 edition of The National Technology Roadmap for Semiconductors[3] projects the advances required in all the industries that support semiconductor manufacturing to maintain the historical rate of improvement (Moore's Law) of integrated circuits. The projections extend to the year 2012, at which time the smallest component of a device would have a linear dimension of 50 nm. However, for years beyond 2006 and device features 100 nm or smaller, the roadmap is filled with the notation "No Known Solution." Indeed, in the Sept. 24 edition of *Science*, Paul Packan of Intel

[3] Semiconductor Industry Association. 1997. *The national technology roadmap for semiconductors: Technology needs*. San Jose, CA (http://www.semichips.org).

described the technical difficulties currently experienced in semiconductor manufacturing. He stated that Moore's Law "now seems to be in serious danger" and that maintaining the rate of improvement in the next decade "will be the most difficult challenge the semiconductor industry has ever faced."[4]

The SIA roadmap ends just short of true nanostructure devices because the principles, fabrication methods, and way of integrating devices into systems are unknown. The roadmap explicitly states that "sustained government support of semiconductor research is mandatory if this industry is to continue to provide for strong economic growth in the U.S." and recognizes that new architectures, materials, and processes will be required to meet the goal of achieving 100 nm feature sizes.

The lead time for science maturing into technology is approximately 10 to 15 years; now is the critical time for Government investment in the science and technology of nanostructures for timely impact in information technology. Further, the investment will have spin-offs that enable the attainment (or acceleration) of other SIA roadmap goals. The area of magnetic information storage is illustrative. Within ten years of the fundamental discovery of the new phenomenon of giant magnetoresistance, this nanotechnology is rapidly replacing older technologies for disk computer heads in a hard disk market worth $34 billion in 1998 (see Figure TS.1).

---

[4] Packan, P.L. 1999. Pushing the limits. *Science* 285: 2079-2081.

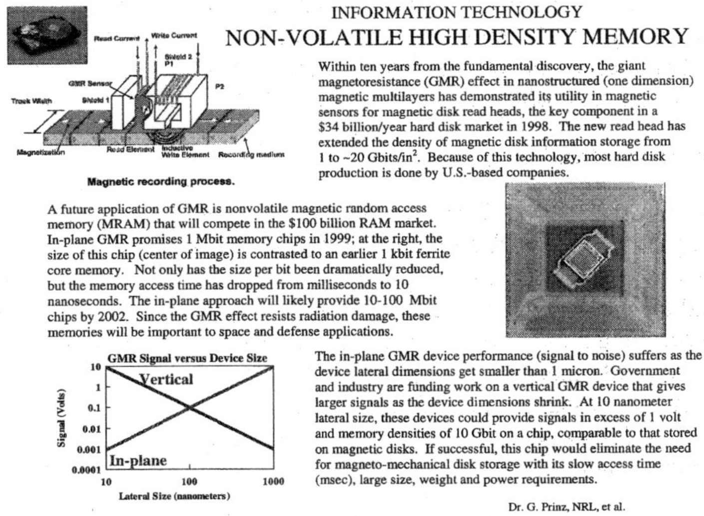

Figure TS.1. Use of a new phenomenon (giant magnetoresistance—GMR) in information technology for non-volatile high density memory.

Other potential nanoelectronics and computer technology breakthroughs include the following:

- Nanostructured microprocessor devices that continue the trend in declining energy use and cost per gate, thereby potentially improving the efficiency of computers by a factor of millions
- Higher transmission frequencies and more efficient utilization of the optical spectrum to provide at least ten times more bandwidth, with consequences in business, education, entertainment, and defense
- Small mass storage devices with capacities at multi-terabit levels, a thousand times better than today
- Integrated nanosensor systems capable of collecting, processing, and communicating massive amounts of data with minimal size, weight, and power consumption

Other potential applications of nanotechnology include affordable virtual reality stations to provide individualized teaching aids (and entertainment); computational capability sufficient to enable unmanned combat and civilian vehicles; and communication capability that obviates much commuting and other business travel in an era of increasingly expensive transport fuels.

Medicine and Health

Recent insights into the uses of nanofabricated devices and systems suggest that today's laborious process of genome sequencing can be made orders of magnitude more efficient through utilization of nanofabricated surfaces and devices. Expanding our ability to characterize an individual's genetic makeup will revolutionize the specificity of diagnostics and therapeutics. Beyond facilitating optimal drug usage, nanotechnology can provide new formulations and routes for drug delivery, enormously broadening the therapeutic potential of such drugs.

Increasing nanotechnological capabilities will also markedly benefit basic studies of cell biology and pathology. As a result of the development of new analytical tools capable of probing the world of the nanometer, it is becoming increasingly possible to characterize the chemical and mechanical properties of cells (including processes such as cell division and locomotion) and to measure properties of single molecules. These capabilities thus complement (and largely supplant) the ensemble average techniques presently used in the life sciences. Moreover, biocompatible, high-performance materials will result from controlling their nanostructure. The molecular building blocks of life—proteins, nucleic acids, lipids, carbohydrates, and their biological mimics—are examples of materials that possess unique properties determined by their size, folding, and patterns at the nanoscale. Based on these biological principles, bio-inspired nanosystems and materials are currently being formed by self-assembly or other patterning methods. Artificial inorganic and organic nanoscale materials can be introduced into cells to play roles in diagnostics (e.g., quantum dots in visualization), but also potentially as active components.

Nanotechnology-enabled increases in computational power will permit the characterization of macromolecular networks in realistic environments. Such simulations will be essential elements in the development of biocompatible implants and in the drug discovery process. There are numerous other potential applications of nanoscience to biology:

- Rapid, efficient genome sequencing, revolutionizing diagnostics and therapeutics
- Effective and less expensive healthcare using remote and in-vivo devices
- New formulations and routes for drug delivery that enormously broaden their therapeutic potential by effecting delivery of new types of medicine to previously inaccessible sites in the body
- More durable, rejection-resistant artificial tissues and organs
- Sensor systems that detect emerging disease in the body, which will shift the focus of patient care from disease treatment to early detection and prevention

Aeronautics and Space Exploration

The stringent fuel constraints for lifting payloads into earth orbit and beyond, and the desire to send spacecraft away from the sun (diminishing solar power) for extended missions, compel continued reduction in size, weight, and power consumption of payloads. Nanostructured materials and devices promise solutions to these challenges. Nanostructuring is also critical to the design and manufacture of lightweight, high-strength, thermally stable materials for aircraft, rockets, space stations, and planetary/solar exploratory platforms. Moreover, the low gravity, high vacuum space environment may aid development of nanostructures and nanoscale systems that cannot be created on Earth. Applications of nanotechnology in this area are broad, with potential relevance to other fields as well:

- Low-power, radiation-tolerant, high-performance computers
- Nanoinstrumentation for microspacecraft
- Avionics made possible by nanostructured sensors and nanoelectronics
- Thermal barrier and wear-resistant nanostructured coatings

Environment and Energy

Nanotechnology has the potential to significantly impact energy efficiency, storage, and production. It can be used to monitor and remediate environmental problems; curb emissions from a wide range of sources; and develop new, "green" processing technologies that minimize the generation of undesirable by-product effluents. The impact on industrial control, manufacturing, and processing will be impressive and result in energy savings. Several technologies that utilize the power of nanostructuring, but that were developed without benefit of the new nanoscale analytical capabilities, illustrate this potential:

- A long-term research program in the chemical industry on the use of crystalline materials as catalyst supports has yielded catalysts with well-defined pore sizes in the range of 1 nm; their use is now the basis of an industry that exceeds $30 billion/year.
- The discovery of the ordered mesoporous material MCM-41 produced by Mobil Oil Co., with pore size in the range 10 to 100 nm, is now widely applied in removal of ultrafine contaminants.
- Several chemical manufacturing companies are developing a nanoparticle-reinforced polymeric material that can replace structural metallic components in the auto industry. Widespread use of those nanocomposites could lead to a reduction of 1.5 billion liters of gasoline consumption over the life of one year's fleet of vehicles and reduce related carbon dioxide emissions by more than five billion kilograms annually.
- The replacement of carbon black in tires by nanometer-scale particles of inorganic clays and polymers is a new technology that is leading to the production of environmentally friendly, wear-resistant tires.

Potential future breakthroughs also include use of nanostructured materials for environmental and nuclear waste management.

National Security

The Department of Defense recognized the importance of nanostructures over a decade ago and has played a significant role in nurturing the field. Critical defense applications of nanotechnology include the following:

- Continued information dominance (see nanoelectronics and computer technology section, pp. x-xii), identified as an important capability for the military
- More sophisticated virtual reality systems based on nanostructured electronics, leading to more affordable, effective training
- Increased use of enhanced automation and robotics to offset reductions in military manpower, reduce risks to troops, and improve vehicle performance; for example, several thousand pounds could be stripped from a pilotless fighter aircraft, resulting in longer missions, and fighter agility could be dramatically improved without the necessity to limit g-forces on the pilot, thus increasing combat effectiveness
- Achievement of the higher performance (lighter weight, higher strength) needed in military platforms, while simultaneously providing diminished failure rates and lower life-cycle costs
- Badly needed improvements in chemical/biological/nuclear sensing and in casualty care
- Design improvement of systems used for nuclear non-proliferation monitoring and management

Other Government Applications

Potential benefits from nanoscience and technology affect other Government missions, including the following:

- Lighter and safer equipment in transportation systems (Department of Transportation)
- Measurement, control, and remediation of contaminants (Environmental Protection Agency)
- Enhanced forensic research (Department of Justice)
- Printing and/or engraving of high quality, forgery-proof documents and currency (Bureau of Engraving and Printing)

Science and Education

The science, engineering, and technology of nanostructures will require and enable advances in a fabric of disciplines: physics, chemistry, biology, materials, mathematics, and engineering. In their evolution as disciplines, they all find themselves simultaneously ready to address nanostructures; this provides a fortuitous opportunity to revitalize their interconnections. The dynamics of interdisciplinary nanoscience efforts will reinforce educational connections between disciplines and give birth to new fields that are unknown at this moment. Further development of the field requires changes in the laboratory and human resource infrastructure in universities and in the education of nanotechnology professionals, especially for industrial careers.

Global Trade and Competitiveness

Technology is the major driving factor for growth at every level of the U.S. economy. Nanotechnology is expected to be pervasive in its applications across nearly all technologies. Investment in nanotechnology research and development is necessary to maintain and improve our position in the world marketplace. The proposed nanotechnology initiative will allow the development of critical enabling technologies with broad commercial potential, such as nanoelectronics, nanostructured materials and nanoscale-based manufacturing processes. These are necessary for U.S. industry to take advantage of nanotechnology innovations and improve our capability to compete globally.

An Outstanding Opportunity and Urgent Responsibility

The potential indicated above for nanotechnology to transform so many aspects of human existence is almost without precedent. In the last few years, applying fundamental discoveries related to nanotechnologies has already developed multibillion-dollar businesses. These latter include giant magnetoresistance (for hard disks), nanolayers (for data storage and the photographic industry), nanoparticles (for drugs in the pharmaceutical field and colorants in printing), confinement effects (for optoelectronic devices and lasers), nanostructured materials (for nanocomposites and nanophase metals), and chemical and biological detection (for national security and the food industry). Fundamentally novel phenomena and processes have led to new, high-value-added technologies. Investment in enabling basic research and infrastructure for nanoscience and engineering promises extraordinarily high economic and societal returns; it is due primarily to this fact that the need to establish a nationally coordinated nanotechnology initiative is so compelling.

National Perspective

According to the 1998 WTEC report summarizing U.S. activities in nanotechnology (Siegel et al. 1998)[5], Federal Government expenditure for nanotechnology in fiscal year 1997 was approximately $116 million. Nanotechnology as defined in that report only included work to generate and use nanostructures and nanodevices; it did not include the simple observation and description of phenomena at the nanoscale that is part of nanoscience. Utilizing the broader definition, the Federal Government expenditure is estimated to be about $255 million for fiscal year 1999. However, 1999 IWGN workshop contributors concluded that a much greater investment could be utilized effectively to increase the rate of discovery, and in fact, many opportunities are not being pursued because of lack of resources. Currently, only about one-third of high quality academic research proposals are being funded. Doubling current Federal expenditures in fiscal year 2001 would ensure that more of the best ideas are funded, increasing the current rate of scientific breakthroughs and drawing more strong researchers to enrich the field. Private industry cannot be expected to fund advancements in basic knowledge on a significant scale. Once nanotechnology has been firmly established, the Government investment will be dwarfed by industry R&D investment, which in the high-technology areas generally is about 10% of sales. Until that time, Government agencies should stimulate and support basic research and infrastructures that will enable subsequent development and commercialization.

---

[5] Siegel, R.W., E. Hu, M.C. Roco, eds. 1998. *R&D status and trends in nanoparticles, nanostructured materials, and nanodevices in the United States.* Baltimore: International Technology Research Institute, World Technology (WTEC) Division, Loyola College. NTIS #PB98-117914. http://itri.loyola.edu/nano/US.Review/.

Nanoscience research in the United States has developed in open competition with existing disciplines. This has been healthy for the early stages of development, but it is also the main reason that U.S. nanotechnology research efforts tend to be fragmented and overlap among areas of relevance and sources of funding. A coordinated effort should focus resources on enabling nanoscience and engineering, on developing infrastructure, stimulating cooperation, and avoiding unwanted duplication of efforts. It should take full advantage of the extraordinarily rich research opportunities and potential technological advances promised by early nanoscience work. A key feature of the IWGN proposal for a national nanotechnology initiative is promotion of synergistic efforts in research, development, and education among Federal agencies.

It is the consensus of the IWGN participants and contributors that the promises of nanotechnology can best be realized through long term and balanced investment in U.S. infrastructure and human resources in five R&D categories in particular:

- Nanostructure properties: Investigate biological, chemical, electronic, magnetic, optical, and structural properties in nanostructures.

- Synthesis and processing: Enable atomic and molecular control of material building blocks to provide the means to assemble and utilize these tailored building blocks for new processes and devices in a wide variety of applications. Extend the traditional approaches to patterning and microfabrication to include parallel processing with proximal probes, stamping, and embossing. Give particular attention to the interface with bionanostructures and bio-inspired structures, to multifunctional and adaptive nanostructures, to scaling approaches, and to affordability at commercial scales.

- Characterization and manipulation: Develop new experimental tools to broaden the capability to measure and control nanostructured matter, including developing new standards of measurement. Pay particular attention to tools capable of measuring and/or manipulating single macro- and supra-molecules of biological interest.

- Modeling and simulation: Accelerate the application of novel concepts and high-performance computation to the prediction of nanostructured properties, phenomena, and processes.

- Device and system concepts: Stimulate the innovative application of nanostructure properties to new technologies.

International Perspective

The United States does not dominate nanotechnology research. There is strong international interest, with nearly twice as much ongoing research overseas as here (see the worldwide study *Nanostructure Science and Technology*, Siegel et al. 1999, NSTC Report[6]). Other regions, particularly Japan and Europe, are supporting work equal to the quality and breadth of the science done in the United States, because there too, scientists and national leaders have determined that nanotechnology has the potential to be a major economic factor during the next several decades. This situation is unlike other post-World War II technological revolutions, where the United States enjoyed earlier leads. Since it will be impossible to lead in every aspect of this emerging super-field, the United States should look to partner with other countries through mutually beneficial information sharing, cooperative research, and study by young U.S. researchers at foreign centers of excellence. We should also build suitable infrastructures to both compete and collaborate with international nanotechnology efforts.

High-Level Recognition of Nanotechnology's Potential

The promise of nanoscience and engineering has not passed unnoticed. Dr. Neal Lane, currently the President's Advisor for Science and Technology and former NSF director, stated at a Congressional hearing in April 1998, "If I were asked for an area of science and engineering that will most likely produce the breakthroughs of tomorrow, I would point to nanoscale science and

---

[6] Siegel, R.W., E. Hu, and M.C. Roco, eds. 1999. NSTC (National Science and Technology Council) Report. *Nanostructure science and technology*. Baltimore: International Technology Research Institute, World Technology (WTEC) Division. Web site: http://itri.loyola.edu/nano/IWGN.Worldwide.Study/. Also published by Kluwer Academic Publishers (1999).

engineering." In March 1998, the President's Science Advisor Dr. John H. Gibbons identified nanotechnology as one of the six technologies that will determine economical development in the next century. NSF started the initiative, Synthesis and Processing of Nanoparticles, in 1991 and the National Nanofabrication User Network in 1994, and has highlighted nanoscale science and engineering in its fiscal year 1998 budget. The Department of Defense identified nanotechnology as a strategic research objective in 1997. The National Institutes of Health identified nanobiotechnology as a topic of interest in its 1999 Bioengineering Consortium (BECON) program.

More recently, on May 12, 1999, Richard Smalley, Nobel Laureate, concluded in his testimony to the Senate Subcommittee on Science, Technology, and Space that "We are about to be able to build things that work on the smallest possible length scales. It is in our Nation's best interest to move boldly into this new field." On June 22, 1999, the House Subcommittee on Basic Research of the Committee on Science organized a hearing on "Nanotechnology: The State of Nano-Science and Its Prospects for the Next Decade." The Subcommittee Chairman Nick Smith, Michigan, concluded the hearings stating that "Nanotechnology holds promise for breakthroughs in health, manufacturing, agriculture, energy use and national security. It is sufficient information to aggressively address funding of this field."

Vision of the Future

Nanoscience and engineering knowledge is exploding worldwide, leading to fundamental scientific breakthroughs and technological paradigm changes in the ways materials, devices, and systems are understood and created. Potential breakthroughs include emergence of entirely new phenomena in physics and chemistry; nanofabrication of three-dimensional molecular architectures; achievement of orders-of-magnitude increases in computer efficiency; utilization of novel data processing architectures such as quantum computing and cellular automata; repair of human tissues with tissue replacements; and realization of a continuous presence in space. In education, nanoscience offers an opportunity to energize the interdisciplinary connections between biology, chemistry, engineering, materials, mathematics, and physics. Nanotechnology will give birth to new fields that at present are only visions of leading researchers.

The national nanotechnology initiative proposed in this report would leverage the existing strong foundation of nanoscience in the United States and address the formidable challenges that remain. It will seize nascent opportunities to advance this field, stimulating domestic job growth and strengthening U.S. competitiveness in international markets. Nanoscale science and engineering promises to become a strategic, dominant technology in the next 10-20 years, because control of matter at the nanoscale underpins innovation and progress in most industries, in the economy, in health and environmental management, in quality of life, and in national security. The consensus of IWGN workshop participants and contributors is that nanotechnology will lead to the next industrial revolution.

RECOMMENDATIONS: A NATIONAL INITIATIVE

There are three fundamental reasons why IWGN workshop participants and contributors believe the time is right for the nation to establish a significant R&D initiative to support nanotechnology: (1) nanotechnology R&D has reached a high level of competitiveness and dynamism, with unusually high, cross-cutting challenges; (2) it is apparent that contributions are necessary from all segments of the science and technology community in order to realize the full potential of nanotechnology R&D; and (3) society's potential return on investment in nanotechnology R&D is immense and of strategic importance. This report proposes a national nanotechnology initiative and outlines its major recommended features. The proposed initiative builds on previous and current nanotechnology programs, including early NSF initiatives on nanoparticles, specialized instrumentation, and functional nanostructures; Department of Defense (DOD) programs supporting its Nanoscience Strategic Research Objectives; and other targeted nanoscience programs of the Department of Energy (DOE), the Department of Commerce (DOC), and the National Institutes of Health (NIH). IWGN workshop participants and contributors addressed the roles that academe, the private sector, the U.S. Government, and professional societies should play in this national nanotechnology initiative.

1. Academe

Role. Universities will continue to play a key role in the development of nanoscience and technology. If there is one signature characteristic of nanoscience, it is its highly transdisciplinary character. This poses difficulties for universities, which mainly are structured in traditional departments. Every effort must be made to foster multidepartmental centers for nanotechnology on campuses. The most successful research efforts will be those that can create new infrastructure (for example, materials preparation and characterization facilities) for these centers.

Recommendations

- Promote interdisciplinary work involving multiple departments.
- Foster on-campus nanotechnology centers for greater interaction.

- Develop new educational paradigms. Introduce nanoscience and engineering in existing and new courses. Include courses on surface science, molecular dynamics, quantum effects, and manufacturing at the molecular scale in curricula at both the undergraduate and graduate levels. Take an integrative science and engineering approach; technology programs cannot be developed without strong supporting science programs because of the scale and complexity of the nanosystems.
- Create or connect "regional coalitions" that involve industry/technology generation.
- Ease intellectual property restrictions to improve information flow with industry.
- Establish graduate and postdoctoral fellowships for interdisciplinary work.

2. Private Sector

Role. The major commercialization opportunities from nanotechnology are probably 10 to 15 years in the future. Much about nanostructures and nanoprocesses is not yet fully measurable, replicable, or understood, and it will require many years to develop corresponding technologies. Industry will invest heavily in nanotechnology only when the underlying capabilities have been developed to the point that products can be foreseen within 3-5 years. Although fundamental nanotechnology research may not be supported by private industry because of the inability of individual companies to restrict and capitalize on basic knowledge, the potential technological and economic benefits that could flow from basic knowledge in nanotechnology are too large for the private sector to ignore. There are critical areas of research and development that can be guided by small industrial teams working within larger consortia of researchers from universities, national laboratories, and other industries. These consortia, perhaps along the lines of the Semiconductor Research Corporation, can provide the critical mass to which companies can afford to contribute. Those that do participate will be in the best position to capitalize on the total research effort and be the first to reach the marketplace with new products derived from nanotechnology.

Recommendations

- Build up investment by maintaining in-house research activities in nanotechnology.
- Join, contribute to, or lead regional coalitions with universities, Government laboratories, and other companies for precompetitive nanotechnology research and information dissemination. The resulting regional centers can encourage niches of common interests, e.g., biotechnology and computers, for jointly developing technologies unlikely to be developed by any single company.
- Sponsor technology start-ups/spin-offs.
- In general, buy into the entire field by investing in nanotechnology research in a variety of ways including training and cross-fertilization among industrial areas.

3. Government R&D Laboratories

Role. Government laboratories can provide many of the large-scale facilities and infrastructure required for fundamental research in nanotechnology, and can serve as technology incubators and provide a stable environment for researchers in the field during the incubation process.

Recommendations

- Pursue applications of nanotechnology in support of respective agency missions.
- Join regional coalitions with universities and industry, and cultivate information flow.
- Provide unique measurement and manufacturing capabilities at nanoscale facilities (synchrotrons, microscopy centers, etc.).
- Provide measurement standards for the nanotechnology field.

4. Government Funding Agencies

Role. Investments must be made in the basic science and technologies that will enable scientists and engineers to invent totally new technologies and stimulate U.S. industrial competitiveness in the emerging nanotechnology areas. The Federal Government should invest in the infrastructure necessary for the United States to lead and benefit from the revolution that is coming. It should support expansion of university and Government/national laboratory facilities, help to build the workforce skills necessary to staff future industries based on nanotechnology, encourage cross-disciplinary networks and partnerships, ensure the dissemination of information, and encourage small businesses to exploit commercial opportunities.

Recommendations

- Undertake a national initiative as part of the fiscal year 2001 budget. The initiative, "National Nanotechnology Initiative (NNI) - Leading to a New Industrial Revolution," should approximately double the Federal Government's annual investment in nanoscience, engineering and technology research and development from the approximately $255 million it spent in fiscal year 1999.

- Address the following priority areas for funding in the initiative:

  A. Long-term fundamental nanoscience and engineering research. The goal is to build fundamental understanding and to discover novel phenomena, processes, and tools for nanotechnology. This commitment will lead to potential breakthroughs and accelerated development in areas such as medicine and healthcare, materials and advanced manufacturing, computer technology, environment and energy. It will refocus the Government's investment that led to today's computer technology and biotechnology.

  B. Synthesis and processing "by design" of engineered, nanometer-size, material building blocks and system components, fully exploiting molecular self-assembly concepts. This commitment will generate new classes of high-performance materials, bio-inspired systems, and efficient, affordable manufacturing of high-performance products. Novel properties and phenomena will be enabled as control of structures of atoms, molecules, and clusters becomes possible.

C. Research in nanodevice concepts and systems architecture. The goal is to exploit nano-derived properties in operational systems and combine building-up of molecular structures with ultraminiaturization. New nanodevices will cause orders of magnitude improvements in microprocessors and mass storage, create tiny medical tools that minimize collateral damage, and enable uninhabited defense combat vehicles in fully imaged battlefields. There will be dramatic payback to other programs with national priority in many fields, including information technology, nanobiotechnology, and medical technology.

D. Application of nanostructured materials and systems to manufacturing, power systems, energy, environment, national security, and health. Basic research is needed in advanced dispersions, catalysts, separation methods, and consolidated nanostructures. Also needed are means for increasing the pace of knowledge development and technology transfer.

E. Education and training of a new generation of skilled workers in the multidisciplinary perspectives necessary for rapid progress in nanotechnology.

- Design a balanced investment strategy that supports a mix of research themes and modes of support, emphasize single principal investigators and small interdisciplinary teams (about two-thirds of the funds on average in all themes), but also support approximately ten R&D centers and networks, such as the existing National Nanofabrication Users Network. Specific agencies should develop project- and disciplinary-oriented activities that focus on education and training; modeling, simulation, and computational science and engineering; infrastructure and facilities; development of partnerships between government, industry, and academia; technology transfer; and international collaboration. Vertical integration activities, pursuing concurrently fundamental research, directed research technology development and prototype construction or clinical evaluations in a collaborative setting, should become a priority for Government R&D laboratories and university- or industry-led consortia. Include support in the initiative for a variety of R&D themes and research modes:

- In support of *fundamental research*, fund single investigators and small groups (30% of the additional investment).
- In support of *grand challenges research*, fund interdisciplinary research and education teams, including those in centers and networks, that have major, long-term objectives (30%).
- In support of *centers and networks of excellence*, fund ten centers for about $5 million each for 5 years, with opportunity for one renewal after review (18%).
- In support of *research infrastructure*, fund development of metrology, instrumentation, modeling and simulation, and user facilities (18%).
- In support of *education and training*, fund student fellowships, traineeships, and curriculum development (4%). Support nanoscience and engineering fellowships that are not tied to one discipline.
- In support of *Small Business Innovative Research (SBIR) and Small Technology Transfer Research (STTR)*, fund focused program announcements on nanotechnology.

• Focus the agencies' programs contributing to the initiative as follows:

DOC	NIST, TA	Measurements and standards; industry-led ventures
DOD	lab and acad.	Information technology; high performance materials; chemical/biological detection
DOE	lab and acad.	Energy science; environment; non-proliferation
DOT	lab and acad.	Smart, lightweight, affordable materials
NASA	lab and acad.	Lighter, smaller spacecraft; radiation-hard electronics
NIH	lab and acad.	Therapeutics; diagnostics; biomaterials; miniaturized tools
NSF	acad.	Science and engineering fundamental knowledge; instrumentation; education

- Direct additional funding toward priority infrastructure requirements.

  A major objective is to create a balanced, predictable, strong, and flexible U.S. infrastructure in nanoscale science, engineering, and technology. This kind of infrastructure is required for the nanotechnology initiative to stimulate further rapid growth of the field. Ideas, concepts, and techniques are developing at an exceedingly rapid pace, such that the field needs coordination and focus with a national perspective. Demands are being made on universities and Government to continue to evolve this science and to bring forth the changes in technology that are expected from the field. Even greater demands are on industry to exploit new ideas, protect intellectual property, and develop appropriate products. This field has major transdisciplinary aspects, which are difficult to coordinate. It is imperative to address these kinds of issues; at stake may be the future economic strength, quality of life, and national security of the United States.

  - Provide nanotechnology investigators with ready access to user-friendly, moderately priced analytical tools in order to carry out state-of-the-art research.
  - Help establish centers with multiple grantees or laboratories where more expensive analytical tools can be made available. These centers should also sponsor the diverse research teams that will be effective in different scientific disciplines. Also, consider ideas concerning remote access and use of these facilities.
  - Use university grants to encourage work among research groups to make maximum use of concepts and ideas being developed in other disciplines. It will be necessary to fund training and fellowships that will attract top quality students. Attention should also be paid to the open exchange of information in multidisciplinary meetings and through rapid publication of research results.
  - Support efforts that will inspire high school students to consider careers in science and engineering and specifically in nanotechnology.

- Promote international collaborations for cost-sharing and joint centers/networks of excellence, where appropriate, for fundamental studies.

- Provide national leadership.

  The Federal Government should provide leadership and maintain coordination and cooperation through an interagency working group to review research thrusts at least annually and promote cooperative efforts. The rapid pace of advances in the field makes this a necessity. This action by the Government will also assist in reducing unwanted redundancy and will make maximum use of appropriated funds.

5. Professional Societies

Role. The science/engineering/technology of nanostructures will flourish best in an interdisciplinary environment with a liberal mix of government, academic and industrial researchers. Professional science and engineering societies must consciously seek to develop appropriate forums that reach beyond their traditional membership and encourage the desired mixing. The societies must also reach out into the international community to assure that U.S. researchers are aware of global advances in nanotechnology.

Recommendations

- Establish interdisciplinary fora that effectively mix academic, government, and industrial researchers, that accelerate progress in research and development in nanoscience, engineering and technology, and facilitate the transition into other fields and technologies.
- Convene groups of scientists and engineers who have not collaborated traditionally.
- Reach out into the international communities to help ensure that worldwide science/engineering advances in nanotechnology are known to the U.S. community.

- Develop symposia to explore educational opportunities at K-12, undergraduate, and graduate levels.
- Invite industrial players to interview prospective scientists and engineers for nano-related openings.

On behalf of the
Interagency Working Group on Nanoscience, Engineering and Technology (IWGN),

Dr. M.C. Roco, National Science Foundation, Chair of the IWGN

Dr. R.S. Williams, Hewlett-Packard Co. (representing the private sector)

Dr. P. Alivisatos, University of California, Berkeley (representing academe)

INTRODUCTION TO NANOTECHNOLOGY FOR NONSPECIALISTS

Contact persons: P. Alivisatos, U.C.-Berkeley; M.C. Roco, NSF; R.S. Williams, H.P.

What is nanotechnology?

Nanotechnology is (1) the creation of useful materials, devices, and systems through the control of matter on the nanometer-length scale, and (2) the exploitation of novel properties and phenomena developed at that scale.

What is a nanometer?

A nanometer is one billionth of a meter (10^{-9} m). This is roughly four times the diameter of an individual atom. A cube 2.5 nanometers on a side would contain about a thousand atoms. The smallest feature in an integrated circuit of today is 250 nanometers on a side and contains about one million atoms in a square layer of atomic height. Proteins, the molecules that catalyze chemical transformations in cells, are 1 to 20 nanometers in size. For comparison, a typical nanometer-scale feature size of about 10 nanometers is 1,000 times smaller than the diameter of a human hair.

Why is this length scale so important?

The wave-like (quantum mechanical) properties of electrons inside matter and atomic interactions are influenced by material variations on the nanometer scale. By creating nanometer-scale structures, it is possible to control fundamental properties of materials like their melting temperature, magnetic properties, charge capacity, and even their color, without changing the materials' chemical composition. Utilizing this potential will lead to new, high-performance products and technologies that were not possible before.

Systematic organization of matter on the nanometer length scale is a key feature of biological systems. Nanotechnology will allow us to place components and assemblies inside cells and to make new materials using the self-assembly methods of nature. In self-assembly, the information necessary for assembly is on the surface of the assembling nanocomponents. No robots or devices are needed to put the components together. This powerful combination of materials science and biotechnology will lead to entirely new processes and industries.

Nanoscale structures such as nanoparticles and nanolayers have very high surface-to-volume ratios, making them ideal for use in composite materials, chemical reactions, drug delivery, and energy storage. Nanostructured ceramics are often both harder and less brittle than the same materials made on the scale of microns, which are 1,000 times larger than nanometers, but still just barely visible to the human eye. Nanoscale catalysis will increase the efficiency of chemical reactions and combustion, at the same time significantly reducing waste and pollution. More than half of therapeutically useful new medicines are not water soluble in the form of micron-scale particles, but they probably will dissolve in water if they are nanometer sized; thus nanostructuring greatly increases the chances of finding new drugs that can be rendered in usable forms.

Since nanostructures are so small, they can be used to build systems that contain a much higher density of components than micron-scale objects. Also, electrons will require much less time to move between components. Thus, new electronic device concepts, smaller and faster circuits, more sophisticated functions, and greatly reduced power consumption can all be achieved simultaneously by controlling nanostructure interactions and complexity.

These are just a few of the benefits and advantages of structuring materials at the nanometer scale.

Is this really new? Don't existing materials already use the nanometer-length scale?

Many existing technologies do already depend on nanoscale processes. Photography and catalysis are two examples of "old" nanotechnologies that were developed empirically in an earlier period despite their developers' limited abilities to probe and control matter at the nanoscale. These two technologies stand to be improved vastly as nanotechnology advances. Most currently existing technologies utilizing nanometer-scale objects were discovered by serendipity, and for many, the role that the nanometer scale played was not even appreciated until recently. For instance, we know now that adding certain inorganic clays to rubber dramatically improves the lifetime and wear properties of tires because the nanometer-sized clay particles bind to the ends of the polymer molecules, which are "molecular strings," and prevent them from unraveling. This is a simple process, but the dramatic improvement in the properties of this composite material, part rubber and part clay, demonstrates the great potential of nanotechnology as it is rationally applied to more complex systems. An example of such a system would be a structure designed to be extremely hard but not brittle, capable of self-repair if minor cracks appear, and easily broken down into its component parts when it is time to recycle the materials.

The ability to specifically analyze, organize, and control matter on many length scales simultaneously has only been possible for about the past ten years. For over a century, chemists have had the ability to control the arrangement of small numbers of atoms inside molecules, that is, to synthesize certain molecules with length scales of less than 1.5 nanometers. This has led to revolutions in drug design, plastics, and many other areas. Over the last several decades, photolithographic patterning (the primary manufacturing process of the semiconductor industry) of matter on the micron length scale has led to the revolution in microelectronics. With nanotechnology, it is just becoming possible to bridge the gap between atom/molecular length scale and microtechnology, and to control matter on every important length scale, enabling tremendous new power in materials design. It is important to remember that the most complex arrangements of matter known to us, living organisms, require specific patterning of matter on the molecular, nanometer, micron, millimeter, and meter scale all at once.

By tailoring the structure of materials at the nanoscale, it is possible to systematically and significantly change specific properties at larger scales—to engineer material behavior. Larger systems constructed of nanometer-scale components can have entirely new properties never before identified in nature. It is also possible to produce composites, i.e., mixtures of different nanoscale entities, that combine the most desirable properties of different materials to obtain characteristics that are greatly improved over those supplied by nature or that appear in combinations not produced by nature. Thus, nanotechnology encompasses a revolutionary set of principles, tools, and processes that will eventually become the foundation for such currently disparate applications as inks and dyes, protective coatings, medicines, electronics, energy storage and use, structural materials, and many others that we cannot even anticipate.

What will be the benefits of nanotechnology?

The new concepts of nanotechnology are so broad and pervasive that they may be expected to influence science and technology in ways that are unpredictable. We are just now seeing the tip of the iceberg in terms of the benefits that nanostructuring can bring (Figure I.1). Existing products of nanotechnologies include wear-resistant tires made by combining nanometer-scale particles of inorganic clays with polymers; nanoparticle medicines with vastly improved delivery and control characteristics; greatly improved printing brought about by utilizing nanometer-scale particles with the best properties of both dyes and pigments; and vastly improved lasers and magnetic disk heads made by precisely controlling layer thicknesses. Many other applications are already under development or anticipated, including those listed below.

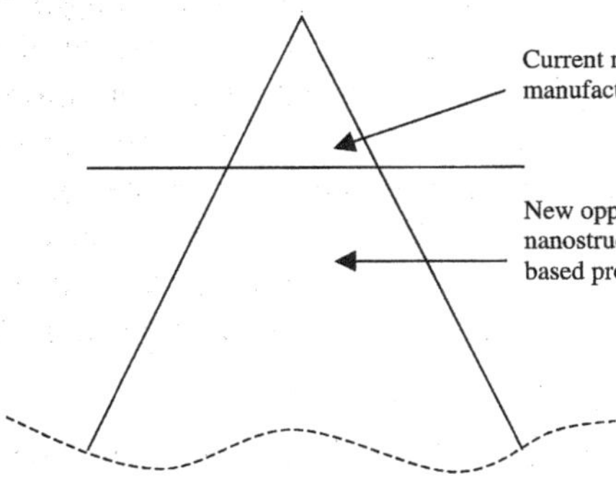

Figure I.1. Current nanotechnology-related materials and manufacturing opportunities.

- *Automotive and aeronautics industries:* nanoparticle-reinforced materials for lighter bodies, nanoparticle-reinforced tires that wear better and are recyclable, external painting that does not need washing, cheap non-flammable plastics, electronics for controls, self-repairing coatings and textiles

- *Electronics and communications:* all-media recording using nanolayers and dots, flat panel displays, wireless technology, new devices and processes across the entire range of communication and information technologies, factors of thousands to millions improvements in both data storage capacity and processing speeds—and at lower cost and improved power efficiency compared to present electronic circuits
- *Chemicals and materials:* catalysts that increase the energy efficiency of chemical plants and improve the combustion efficiency (thus lowering pollution emission) of motor vehicles, super-hard and tough (i.e., not brittle) drill bits and cutting tools, "smart" magnetic fluids for vacuum seals and lubricants
- *Pharmaceuticals, healthcare, and life sciences:* new nanostructured drugs, gene and drug delivery systems targeted to specific sites in the body, biocompatible replacements for body parts and fluids, self-diagnostics for use in the home, sensors for labs-on-a-chip, material for bone and tissue regeneration
- *Manufacturing:* precision engineering based on new generations of microscopes and measuring techniques, new processes and tools to manipulate matter at the atomic level, nanopowders that are sintered into bulk materials with special properties that may include sensors to detect incipient failures and actuators to repair problems, chemical-mechanical polishing with nanoparticles, self-assembling of structures from molecules, bio-inspired materials and biostructures
- *Energy technologies:* new types of batteries, artificial photosynthesis for clean energy, quantum well solar cells, safe storage of hydrogen for use as a clean fuel, energy savings from using lighter materials and smaller circuits
- *Space exploration:* lightweight space vehicles, economic energy generation and management, ultra-small and capable robotic systems
- *Environment:* selective membranes that can filter contaminants or even salt from water, nanostructured traps for removing pollutants

from industrial effluents, characterization of the effects of nanostructures in the environment, maintenance of industrial sustainability by significant reductions in materials and energy use, reduced sources of pollution, increased opportunities for recycling

- *National security:* Detectors and detoxifiers of chemical and biological agents, dramatically more capable electronic circuits, hard nanostructured coatings and materials, camouflage materials, light and self-repairing textiles, blood replacement, miniaturized surveillance systems

Additional details on ongoing nanotechnology R&D results may be found in chapters 4-10 of this report. A modified Delphi survey of experts on nanotechnology taken in May 1999 suggests that the probability of commercial applications of these and other nanotechnology processes and products in the next 15-25 years is between 50 and 100 percent, with the majority of applications ranging from 90 to 100 percent probability of commercialization (Gutmanis 1999).

What should Government do to ensure the United States can enjoy the envisioned benefits?

Government can play the key role to assure that the United States realizes the enormous benefits of nanotechnology. The goals of nanotechnology research are too fundamental, long-term (greater than ten years), transdisciplinary, and high-risk for industry to take an immediate leadership role, although there is high level of industry interest. Given the expectations of U.S. investors and the competitiveness of the global marketplace, U.S. industry is unable to invest significantly in long-term and thus risky research that takes many years to develop into products. In the United States, the university and government research systems must fill this gap.

Because of its transdisciplinary nature, nanotechnology will require teams of physicists, chemists, biologists, and engineers to develop its viability as a field. Government agencies will need to foster this teamwork. A worldwide competition is already underway in this area, and the U.S. response to date is fragmented in comparison to the approaches of European and Asian countries (see Siegel et al. 1999,

NSTC report). Moreover, new infrastructure at universities and the national labs is needed for the field to develop. The increasing pace of technological innovation and commercialization demands continual compression of the discovery-invention-development time scales, which in turn requires parallel and coordinated work in both basic research and commercial product development. The requirements for and from nanotechnology transcend anything that can be supplied by traditional academic disciplines, national laboratories, or even entire industries. For all of these reasons, a Federal initiative is critical to establishing an effective national effort in nanotechnology.

Looking to the future: lessons from the past

Although there is always considerable uncertainty in predicting future benefits, this report attempts to anticipate some that will occur within the next few decades. A significant lesson of the 20th century is that predictions of the state of a particular technology several decades in the future often fall far short of what is actually accomplished. One famous example of such a prediction was that made in the March 1949 edition of *Popular Mechanics*, in which several experts confidently predicted that the computers of the future would add as many as 5,000 numbers per second, weigh only 3,000 pounds, and consume only 10 kilowatts of power. Although at the time this was a bold forecast, it seems quaint now, when there are laptop computers that can add several million numbers per second while using only about a watt of power. Another famous prediction from the 1950s was that the total world market for electronic computers would be fewer than 10, whereas now there are about a billion microprocessors operating as the key components of computers, cellular telephones, automobiles, games, medical imaging instruments, and many more applications. The computer industry is one of the largest and healthiest in the United States, providing a substantial volume of exports and high-paying jobs. It has also spawned other enormous and important industries, such as computer software, that were not even envisioned fifty years ago.

The reason that the sages of *Popular Mechanics* could not foresee the advent of the information industry was that they anticipated only <u>evolutionary</u> change. Their predictions for the future of computers would probably have been correct if computers were still built with vacuum tubes and relays. However, a technological revolution was already beginning: the transistor had been invented in 1947. Along with the integrated circuits that appeared a decade later, this discovery ushered in a new industrial era, the age of silicon and information. The new epoch was born and nurtured in the United States because of the broad fundamental and applied research base that existed here at that period and the sustained Federal investments that went into training the people, building the scientific infrastructure, and creating the culture in which ideas could flow within a broad community of scientists and engineers from academe and industry. One can only speculate what life would be like in the United States today if this technological revolution had originated in a different country or if it had not occurred at all.

The total societal impact of nanotechnology is expected to be much greater than that of the silicon integrated circuit, because it is applicable to many more fields than just electronics. Significant product performance improvements and manufacturing advances will lead to many industrial revolutions in the twenty-first century. Nanotechnology has the potential to change the nature of almost every human-made object, because control at the nanoscale means tailoring the fundamental properties, phenomena, and processes exactly at the scale where electronic, chemical, and biological properties and phenomena are defined. A major question is how can we embrace and facilitate the nanotechnology revolution to maximize the benefit to all U.S. citizens.

References

Gutmanis, Ivars. 1999. *Probability of the nanotechnology manufacturing processes in the industrial nations in 2015-2025 time period.* Report of Hobe Corporation.

Siegel, R.W., E. Hu, and M.C. Roco, eds. 1999. NSTC (National Science and Technology Council) Report. *Nanostructure science and technology.* Baltimore: International Technology Research Institute, World Technology (WTEC) Division. Web site:
http://itri.loyola.edu/nano/IWGN.Worldwide.Study/.
Also published by Kluwer Academic Publishers (1999).

Chapter 1

FUNDAMENTAL SCIENTIFIC ISSUES FOR NANOTECHNOLOGY

Contact persons: G. Whitesides, Harvard University; P. Alivisatos, U. California, Berkeley

1.1 VISION

Nanostructures are the entry into a new realm in physical and biological science. They are intermediate in size between molecular and microscopic (micron-size) structures. They contain a countable number of atoms, and are, as a result, uniquely suited for detailed atomic-level engineering. They are chameleons: viewed as molecules, they are so large that they provide access to realms of quantum behavior that are not otherwise accessible; viewed as materials, they are so small that they exhibit characteristics that are not observed in larger (even 0.1 μm) structures. They combine small size, complex organizational patterns, potential for very high packing densities and strong lateral interactions, and high ratios of surface area to volume.

Nanostructures are the natural home of engineered quantum effects. Microstructures have formed the basis for the technologies that support current microelectronics. Although microstructures are small on the scale of direct human experience, their physics is largely that of macroscopic systems. Nanostructures are fundamentally different: their characteristics—especially their electronic and magnetic characteristics—are often dominated by quantum behavior. They have the potential to be key components in information technology devices that have unprecedented functions. They can be fabricated in materials that are central to electronics, magnetics, and optics. Nanostructures are, in a sense, a unique state of matter—one with particular promise for new and potentially very useful products.

Because they are small, nanostructures can be packed very closely together. Their high packing density has the potential to bring higher speed to information processing and higher areal and volumetric capacity to information storage. Such dense packing also is the cause of complex electronic and magnetic interactions between adjacent (and sometimes, nonadjacent) structures. For many nanostructures, especially large organic molecules, the small energetic differences between their various possible configurations may be significantly shaped by those interactions. In some cases, the presence of surface interface material, with properties different from the nanostructures themselves, adds another level of complexity. These complexities are completely unexplored, and building technologies based on nanostructures will require in-depth understanding of the underlying fundamental science. These complexities also promise access to complex non-linear systems that may exhibit classes of behavior fundamentally different from those of both molecular and microscale structures.

Exploring the science of nanostructures has become, in just a few years, a new theme common to many established disciplines. In electronics, nanostructures represent the limiting extension of Moore's law and classical devices to small devices, and they represent the step into quantum devices and fundamentally new processor architectures. In molecular biology, nanostructures are the fundamental machines that drive the cell—histones and proteosomes—and they are components of the mitochondrion, the chloroplast, the ribosome, and the replication and transcription complexes. In catalysis, nanostructures are the templates and pores of zeolites and other vitally important structures. In materials science, the nanometer length scale is the largest one over which a crystal can be made essentially perfect. The ability to precisely control the arrangements of impurities and defects with respect to each other, and the ability to integrate perfect inorganic and organic nanostructures, holds forth the promise of a completely new generation of advanced composites. Each of these disciplines has evolved its own separate view of nanoscience; the opportunities for integrating these views and for sharing tools and techniques developed separately by each field are today among the most attractive in all of science.

Nanoscience is one of the unexplored frontiers of science. It offers one of the most exciting prospects for technological innovation. And if it lives up to its promise as a generator of technology, it will be at the center of fierce international competition.

1.2 CURRENT SCIENTIFIC AND TECHNOLOGICAL ADVANCEMENTS

Nanoscience has exploded in the last decade, primarily as the result of the development of new tools that have made the characterization and manipulation of nanostructures practical, and also as a result of new methods for preparation of these structures.

Scaling Laws and Size-dependent Properties of Isolated Nanostructures

It is now well established that such fundamental properties as the melting temperature of a metal, the remanence of a magnet, and the band gap of a semiconductor depend strongly upon the size of the component crystals, provided they are in the nanometer regime. Almost any property in a solid is associated with a particular length scale, and below this length, the property will vary. For instance, the exciton diameter in a semiconductor may be tens or hundreds of nanometers, the distance between domain walls in a magnet may be hundreds of nanometers, etc. This opens the prospect for creating a new generation of advanced materials with designed properties, not just by changing the chemical composition of the components, as has been done in the past, but by controlling the size and shape of the components. This creates great opportunities for fundamental science in condensed matter physics, solid state chemistry, materials science, electrical engineering, biology, and other disciplines.

Tools for Characterization

Scanning probe microscopies have revolutionized characterization of nanostructures, and development of new variants of scanning probe devices continues apace. Older tools, especially electron microscopy, continue to play essential roles. In biological nanoscience, the combination of X-ray crystallography and NMR spectroscopy offers atomic-resolution structural information about structures as complex as entire virus particles.

Fabrication and Synthesis

Tremendous advances are currently occurring in the synthesis and fabrication of isolated nanostructures. These activities range from colloidal synthesis of nanocrystals to the growth of epitaxial quantum dots by strained layer growth. Related activities include the preparation of fullerenes, buckytubes, and other one-dimensional nanostructures, as well as the growth of mesoporous inorganics. Increased activity in the nanoscale design of polymers is also occurring, including the development of dendrimers and complex block copolymers. The techniques of molecular biology have made a very wide range of biological nanostructures readily available through cloning and overexpression in bacterial production systems.

While much has been accomplished in the growth of isolated nanostructures, work has only just begun in the use of self-assembly techniques to prepare complex and designed spatial arrangements of nanostructures. A parallel line of current activity in fabrication of patterned nanostructures rests on the extension of techniques highly developed in the field of microelectronics: photolithography, X-ray lithography, and e-beam lithography. A number of recent developments in synthesis and fabrication offer the potential both to generate new types of structures, and, probably more importantly, to generate these structures at a fraction of the cost of techniques derived from microlithography. Soft lithography, which uses molding, printing, and embossing to form patterned structures in plastics and glasses, has expanded the range of materials that can be used and has suggested routes to previously inaccessible three-dimensional structures.

Computation

Because nanostructures contain few atoms (at least relative to most materials), they are uniquely susceptible to high-level simulation using supercomputers. The capability to treat nanostructures with useful accuracy using computation and simulation will be invaluable both in fundamental science and in applied technologies.

Emerging Uses

Many clear applications for nanotools and nanostructures are already evident and are the targets of existing technology development programs:

- Giant magnetoresistance (GMR) materials have been introduced into commercial use with remarkable speed, and their acceptance suggests the importance of magnetic materials with nanometer-scale spin-flip mean free path of electrons.
- Numerous nanodevices and nanosystems for sequencing single molecules of DNA have been proposed; these structures, if successful, will be invaluable in the Human Genome Project and other large-scale genomics programs. Indeed, it seems quite likely that there will be numerous applications of inorganic nanostructures in biology and medicine, as markers.
- Similarly, there exists a range of ideas for high-density information storage, based, for example, on concepts such as nano-CDs and on nanostructured magnetic materials, including materials showing giant and tunneling magnetoresistive effects; these promise to provide future systems for memory with ultrahigh densities.
- New types of components for information processors based on quantum mechanical principles (resonant tunneling transistors; single electron transistors; cellular automata based on quantum dots) are being explored actively at the level of research; these types of processors appear to fit well in the burgeoning field of quantum computation.
- New protective coatings, thin layers for optical filtering and thermal barriers, nanostructured polymers, and catalysts are already coming to the market. Nanostructured coatings are showing good corrosion/erosion resistance as possible replacements for the environmentally troublesome chromium-based coatings.
- Aerogels—highly porous, sponge-like materials with a three-dimensional filigree of nanostructures—have promise in catalysis and energy applications.

1.3 GOALS FOR THE NEXT 5-10 YEARS: BARRIERS AND SOLUTIONS

Numerous important areas require active research and development. The objectives of current research are to be able to understand the properties of isolated nanostructures; to make arbitrary structures with atomic-level precision; to do so rapidly, in large numbers, and inexpensively; and to design these structures to have desired properties using appropriate computer tools. Nanoscience is far from this objective, but it is moving rapidly in every component of the problem.

Fundamental Properties of Isolated Individual Nanostructures

Individual nanostructures in isolation are the building blocks of nanotechnology. Individual structures are studied because we do not yet know the fundamental limits to the preparation of *identical* nanostructures and because each nanostructure can interact differently with its environment. Underlying the fundamental properties of nanostructures are two broad themes. First, the size-dependent properties of materials in the nanoscale regime are predicted to vary qualitatively according to scaling laws; comparison to these simple scaling laws remains an important activity. Second, the properties of isolated nanostructures have a significant statistical variation, fluctuating in time, and it is important to observe and understand these variations.

Many key questions relate to the structure, or arrangement of atoms, in a nanostructure. The relative stability of different structural phases is altered in the nanometer regime, affected by both kinetic and thermodynamic factors. Variations may arise for many reasons, including surface energies, absence of defects, or electronic quantum size effects. There is a compelling need to map out the kinetics and thermodynamics of phase transformations in nanostructures. For any picture of the physical properties of nanostructures to be complete, the structure of the surface must also be determined. Due to their finite sizes, the structure and composition of the surfaces of nanostructures may have particular importance for their chemical and physical properties. The surfaces of nanostructures are likely to vary significantly from the well-known structures of bulk surfaces, and entirely new experimental techniques for measuring these reconstructions need to be developed. The ability not only to measure, but ultimately to systematically control, the surface and interior structures of nanoscale materials will be an ongoing field of research over the next decade.

Recent developments have permitted the observation of optical, electrical, magnetic, chemical, thermal, mechanical and biological properties of isolated, individual nanostructures. These techniques, which have facilitated developments such as single-electron transistors, scanning probe microscopies, and single-molecule spectroscopy, have revolutionized our understanding of nanostructured materials. The early studies point the way to a long-term agenda: develop new probes of fundamental properties that can work with nanometer spatial resolution and ever-improved temporal resolution and sensitivity.

Early work also reveals that nanoscale systems can be fluctuational by nature. Much work is needed to understand their fluctuations and to learn what the fundamental limits are. As an example, an important question in quantum computation concerns whether it is possible to prepare well-defined superposition of quantum states in nanostructures without rapid dephasing.

Fundamental Properties of Ensembles of Isolated Nanostructures

Nanostructures may be used in a wide range of contexts; most of these are ones in which ensembles of nanostructures are assembled into a complex, functional arrangement. Many properties of nanoscale building blocks vitally depend upon the size, the shape, and indeed the precise arrangement of all the atoms within. Thus, a high priority must be placed upon understanding the fundamental limits to the preparation of identical nanostructures. To this day, the processes of nucleation and growth are incompletely understood, and we do not know what is the largest number of atoms that can be assembled into a precisely defined molecular structure. Even if a system is prepared according to an ideal chemical process, there inevitably will be variations. There is a need to understand further which desirable properties are retained or even simplified by averaging out such variations, and which properties are lost.

Assemblies of Nanoscale Building Blocks

The construction of functional assemblies of nanostructures depends upon a sound understanding of the intrinsic couplings between nanostructures. Charge separation and transport, tunneling, through-space electromagnetic coupling, and mechanical and chemical interactions between nanostructures need to be measured and theoretically described further than has been done to date.

A combination of electron beam (and perhaps X-ray) lithography, scanning probe writing and fabrication, soft lithography, self-assembly, and catalytic growth together offer a rich menu of new and old fabrication techniques to use in patterning nanostructures. With the exception of e-beam and X-ray lithography, these techniques are all early in their development cycle and have substantial promise for growth. The next decade will see these techniques developed and integrated into a suite of methods for the fabrication of nanostructures and nanosystems.

The synthesis of bulk materials—colloids, magnetic structures, zeolites, buckytubes and analogs, aerogels, and many others—is also an area where there is potential for great innovation. In most of these materials, the physics underlying their behavior is understood only incompletely; there is an opportunity now to discover many new behaviors resulting from confinement of electrons or photons, from high structural perfection, from high ratios of surface to volume, or from some other aspect of small size.

In relevant but independent research in molecular biology, techniques of genetic manipulation, combined with technologies for protein production, provide the way to make small quantities of almost any protein of interest. The genome projects will extend this capability. Progress in biotechnology will increase the ease with which large quantities of genetic materials can be generated.

A key issue is understanding the integration processes of various isolated nanostructures and assemblies of nanostructures.

Evaluation of Concepts for Devices and Systems

Among the most critical impediments to thinking seriously about nanostructure-based systems are the difficulties in understanding how they are to be interconnected and addressed and in understanding what kinds of new functions are achievable. For instance, in nanoelectronics, there are a number of solutions that have been suggested for some of these issues—ranging from using buckytubes as nanowires to addressing individual components via ganged scanning probe devices, or even optically—but there has been almost no serious work directed toward the problems of systems fabrication. Among the numerous questions that must be solved are: what to use as wires; how to design and fabricate devices; and what the architectures of systems should be.

Research is currently focused on exploring the fabrication and characterization of single devices at the level of single electron transistors or resonant tunneling devices. There is much less effort (and certainly much less than is needed) devoted to asking fundamental questions about how electrical current is to be carried from a contact pad to a device in a densely packed array, fault- and defect-tolerant designs, device isolation, the operation of large arrays of cooperative devices, and other fundamental questions dealing with the problem of making an array of nanostructured quantum devices that performs some complex function. Similar to the nanoelectronics issues, resolving the aspects related to integration at nanoscale is essential in dispersions, nanocomposites, sensors, and other areas.

Nanomanufacturing

Developing techniques for fabricating nanostructures inexpensively in very large numbers—that is, manufacturing them—is an area that requires substantial effort: nanoscience will not be fully successful until it has provided the base for manufacturing technologies that are economically viable. It is probable that methods developed for microfabrication in the >100 nm size range will not work in the 20 nm range. It will thus probably be necessary to develop an entire new suite of manufacturing methods for nanostructures. There is reason to believe that self-assembly and soft lithography will be able to make substantial contributions to this important problem, but other fundamentally new methods will also be needed.

Connecting Nanoscience and Biology

One of the opportunities in basic science is to search for synergies between nanoscience in biology and nanoscience as developed in the contexts of computation and information science and solid-state physics and chemistry. There is no question that understanding the structure and function of biological nanostructures will stimulate fabrication of nonbiological materials; it is possible that biologically derived structures may also be useful in assembly of systems of nanodevices. In return, nanofabrication can provide analytical tools for investigating biomolecules (in genomics, proteomics, and high-throughput screening for drug leads) as well as for exploring the interior structure and function of cells. One objective of nanoscience should be to build robust intellectual bridges between its currently scattered disciplinary components, but especially between nanoelectronics and molecular biology.

Molecular Electronics

Molecular electronics offers an attractive opportunity for basic science in nanosystems. Organic molecules are probably the smallest systems that can be imagined for many possible functions, but they have the disadvantage, from the point of view of possible use in nanoelectronic systems, that they are usually poor conductors of electricity. A number of systems have been investigated in which experimental results suggest that organic molecules can act as molecular wires; there are hints that they may also serve as components in more complex systems. The science base in molecular electronics is very early in its development, and it is not yet clear whether organic molecules—even if they do conduct electricity—have the other properties they would need to be the core components of a large-scale nanoelectronics technology. Defining the real promise of molecular electronics should be an objective of research.

Nanostructures as Model Systems for Earth and Planetary Science

Nanoscale components are fundamental building blocks for solid state chemistry. Thus, further study of nanostructures can lead to improvements in our understanding of fundamental processes in earth and planetary science. Improved understanding of interstellar dust, the formation mechanisms of minerals, and the processes of weathering can all result from fundamental studies of nanostructures.

1.4 SCIENTIFIC AND TECHNOLOGICAL INFRASTRUCTURE

Tools for Synthesis and Fabrication, and for Characterization

To work in nanoscience, it is a prerequisite to be able to fabricate and characterize nanostructures: to make rabbit stew, one must first catch a rabbit. Making the important tools for fabrication and characterization available to user communities is an imperative; throughout the nation, research in nanoscience is still limited by limited access to tools. Certain instruments, especially electron microscopes, are sufficiently expensive that they should be operated within consortia; others, especially state-of-the-art scanning probe devices, should be distributed to qualifying individual research groups. The character and cost of the facilities needed for fabrication generally depend on the technique being used. For the more expensive facilities—for example, high-resolution e-beam writers, good clean-room facilities, and mask-making facilities—a substantial, early investment is needed to prevent fabrication delays.

Flexible Research Structure

Nanoscience is an area in which there is no single way to do research: both single-investigator, peer-reviewed research and programmatic research involving groups of investigators from different disciplines and organizational structures have roles to play. It is important not to narrow the options for modalities of research support at this stage in the development of the field.

Training the Next Generation of Nanoscientists

Nanoscience has emerged into prominence only in the last 5-10 years. It is not supported in universities by existing departments: nanoscientists come from chemistry, physics, biology, electrical engineering, and others, and tend to think of themselves in terms of their historical disciplinary affiliations. Building educational programs that focus on nanoscience and nanotechnology as a distinct field—whether addressing device physics or structural biology—would provide a profound boost to the advancement of nanoscience by making it visible and intellectually attractive to the brightest young people and by promoting transfer of information between disciplines.

1.5 R&D INVESTMENT AND IMPLEMENTATION STRATEGIES

Because nanoscience is an area in which a range of research styles flourish, it is important to keep support distributed among a number of sources; it should not, at this juncture, be concentrated in one Federal agency. The National Science Foundation (NSF) and the Department of Defense Advanced Research Projects Agency (DARPA), for example, serve fundamentally different functions in nanoscience, and consolidation of these functions into one agency could slow progress in the field dramatically. One of the challenges of an area of national importance but in which Government support is distributed to various agencies, is to coordinate Government-sponsored activities so that the whole is greater than the sum of the parts. NSF is the plausible agency to lead the coordination effort.

Reporting

To maintain a high focus on nanoscience/technology and to ensure careful reporting, NSF should be charged with providing an annual appraisal of the field to the Office of Management and Budget (OMB) and other interested parties such as various House and Senate committees, industrial users, etc. Appropriate reporting and interagency discussion would help to maximize productive transfer of processes and materials among groups, regardless of the source of support.

Emphasis in Support

Nanoscience is still in its infancy. The emphasis in Federally funded research should be on precompetitive fundamental science and engineering. The object of the work should be to develop flexible, innovative methods of making, characterizing, applying, and manufacturing nanosystems. In nanoscience, as in many fields, technology transfer represents a difficult step. In other fields of technology—especially biotechnology and information technology—the venture community has played a major role in technology transfer. Developing intellectual property and financial policies that make nanoscience an attractive investment would accelerate the development of commercial technologies.

1.6 PRIORITIES AND CONCLUSIONS

Interest in nanoscience is burgeoning, largely due to its enormous practical potential. Yet it is very clear that nanoscience has provided a new way of thinking that has the potential to promote truly exciting developments in fundamental science as well. Long-standing questions in condensed matter physics and chemistry, in biological science, in materials science, and in mechanical engineering will receive renewed attention as a result of the exciting developments in nanotechnology.

Priorities

a. Build a broad program of R&D in nanoscience that will include research universities, relevant industry, and some of the national laboratories. Important public policy objectives in this field should include building a "community" focused on nanoscience/technology and providing stable support for this community at a level high enough to allow the participating groups to reach a critical mass.

b. Develop other mechanisms for bridging the gaps between communities interested in nanoscience, including both the physical and biological sciences.

c. Develop policies explicitly designed to attract large companies as participants in programs of Federally funded groups. Without the large-company participation, technology development programs in nanoelectronics will probably fail at the stage of research planning and product definition.

d. Develop a strategy for informal coordination of R&D among participating Federal agencies, centered in NSF.

e. Maintain an active series of reports to OMB and Congress, both to aid in educating policymakers about the progress, opportunities, and failures of the field, and to provide a general education about nanoscience at senior levels in the Government.

f. Provide Federal funds to push science that seems to offer the potential of developing into profitable technology rapidly to the point of manufacturable prototypes using focused, DARPA-style programs.

g. Address the problems of public perception of threats from nanoscience with active programs to reduce any possible threats and educate the public.

1.7 EXAMPLES OF CURRENT ACHIEVEMENTS AND PARADIGM SHIFTS

1.7.1 Quantum Dot Formed by Self-assembly (Ge "pyramid")

Contact person: R.S. Williams, Hewlett-Packard Co.

Figure 1.1 is a scanning tunneling microscope (STM) image of a pyramid of germanium atoms on top of a silicon surface. The pyramid is ten nanometers across at the base, and it is actually only 1.5 nanometers tall (the height axis in the image has been stretched to make it easier to see the detail in the faces of the pyramid). Each round-looking object in the image is actually an individual germanium atom.

The pyramid forms itself in just a few seconds in a process called "self-assembly." If the proper number of germanium atoms is deposited onto the correct type of silicon surface, the interactions of the atoms with each other causes the pyramids to form spontaneously. The propensity of some materials to self-assemble into nanostructures is currently a major area of research. The intent is to learn how to guide or modify self-assembly to get materials to form more complex structures, such as electronic circuits.

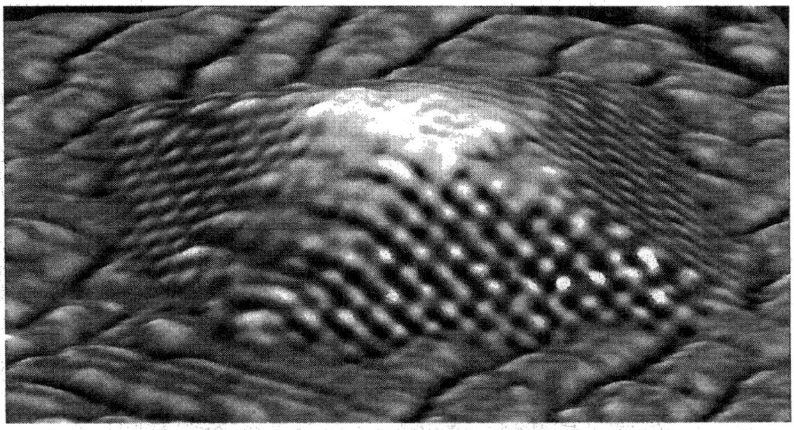

Figure 1.1. STM image of quantum dot formed by self-assembling (Ge "pyramid") (courtesy Hewlett-Packard; image acquired by G. Medeiros-Ribeiro, Hewlett-Packard Labs).

Manufacturing processes based on guided self-assembly of atoms, molecules, and supramolecules promise to be very inexpensive. Instead of requiring a multibillion-dollar manufacturing facility, electronic circuits of the future may be fabricated in a beaker using appropriate chemicals, and yet they may be many thousands to millions of times more capable than current chips. Just twenty years ago, few scientists even dreamed it would be possible to see such a detailed picture of the atomic world, but with the advent of new measuring tools, the scanning probe microscopes developed in the mid-1980s, seeing atoms is now an everyday occurrence in laboratories all over the world.

1.7.2 Quantum Corral

Contact person: D. Eigler, IBM

Figure 1.2 is a scanning tunneling microscope image of a "quantum corral." The corral is formed from 48 iron atoms, each of which was individually placed to form a circle with a 7.3 nanometer radius. The atoms were positioned with the tip of the tunneling microscope. The underlying material is pure copper. On this copper surface there are a group of electrons that are free to move about, forming a so-called "two-dimensional electron gas." When these electrons encounter an iron atom, they are partially reflected. The purpose of the corral is to try to trap, or "corral" some of the electrons into the circular structure, forcing the trapped electrons into "quantum" states. The circular undulations in the interior of the corral are a direct visualization of the spatial distribution of certain quantum states of the corral. Experiments such as this give scientists the ability to study the physics of nanometer-scale structures and to explore the potential application of these small structures to any of a number of purposes.

Figure 1.2. STM image of a "quantum corral" (courtesy IBM Research Division).

1.7.3 Magnetic Behavior at Nanoscale

Contact person: P. Alivisatos, U.C.-Berkeley

Figure 1.3 is a transmission electron microscope (TEM) image of an elegant example of natural nanotechnology that occurs in magnetotactic bacteria. These are bacteria that contain within them a "compass" that allows them to move in a particular magnetic direction. The compass consists of a series of magnetic nanoparticles arranged in a line.

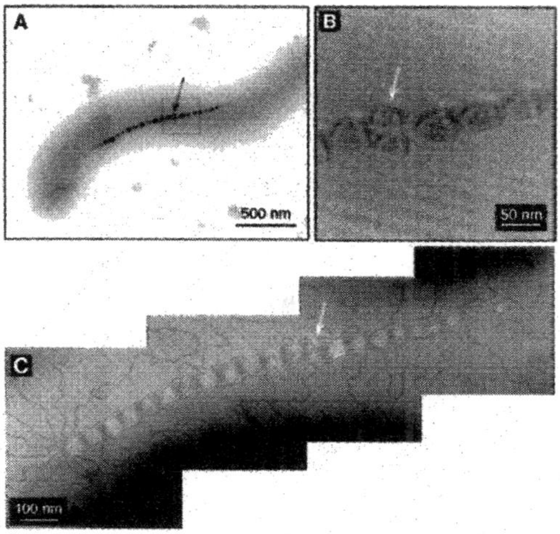

Figure 1.3. TEM image showing natural nanotechnology in magnetotactic bacteria (*magnetospirillum magnetotecticum* strain MS-1). Chains of nanocrystals used for navigation illustrate nature's exploitation of a fundamental scaling to achieve maximum and most efficient use of magnetization (reprinted with permission from Dunin-Borkowski et al. 1998, ©1998 American Association for the Advancement of Science).

Each particle is as large as it can be and still remain a single magnetic domain, 25 nm. Larger particles have a type of defect, a magnetic domain wall, that lowers their coercivity. In the bacteria, these magnetic particles aggregate spontaneously into chains. The resulting compass uses a minimum amount of material to achieve the desired property, alignment along Earth's magnetic field.

Artificial nanotechnology researchers can learn much from what already occurs in nature. The time required for a magnetized specimen to lose memory of its direction of magnetization depends exponentially upon the volume, provided the crystal is still in the nanometer regime. Thus, the same physics imposes a lower limit of a few tens of nanometers to the size of an iron oxide particle that could be used in a magnetic memory unit that operates at room temperature.

1.7.4 Ordered Monolayer of Gold Nanocrystals

Contact person: R. Andres, Purdue University

Figure 1.4 is a transmission electron microscope image of an ordered monolayer of gold particles 5 nm in diameter supported on a thin carbon membrane. The round-looking objects are well-faceted single crystals of gold atoms. These nanocrystals have the shape shown in the bottom right inset and are aligned as a hexagonal array as illustrated by the expanded view shown at the bottom left. Each gold particle is mechanically and electronically separated from its nearest neighbors by organic molecules, which give structural integrity to the monolayer and serve as a controlled tunnel barrier for electron transport between the particles. This ordered monolayer forms spontaneously on a water surface and can be transferred intact to a wide range of flat solid substrates. If instead of transferring the entire monolayer, a method can be developed to transfer only a selected pattern of narrow ribbons, it would provide an elegant solution to the problem of interconnecting electronic devices having smaller and smaller dimensions.

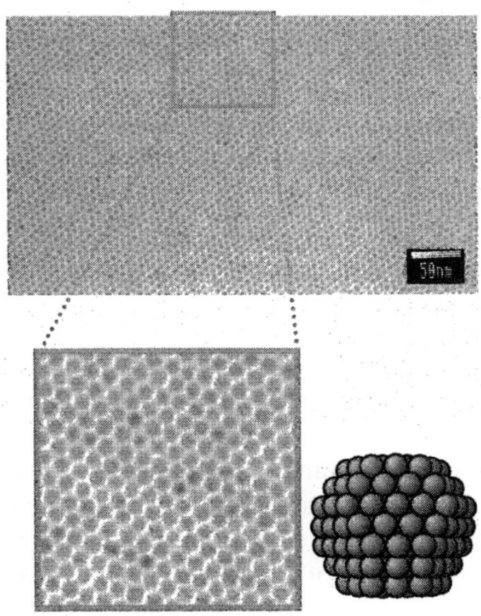

Figure 1.4. TEM image of an ordered monolayer of gold nanocrystals.

1.7.5 Nanostructured Polymers

Contact person: S. Stupp, Northwestern University

Figure 1.5 depicts a supramolecular nanostructure formed by the ordered self-assembly of triblock copolymers. The polar liquid-crystalline parts of the molecules (bottom) arrange themselves in an ordered lattice, while the bulky, aromatic-hydrocarbon units (top) form an amorphous cap. These nanosized mushroom-shaped units further self-assemble into polar sheets whose top surfaces (mushroom caps) are hydrophobic and whose bottom surfaces (mushroom stems) are hydrophilic. Such self-assembled nanostructures are under investigation as anti-icing coatings for aircraft, lubricating layers for microelectronics, anti-thrombotic agents for arteries, etc.

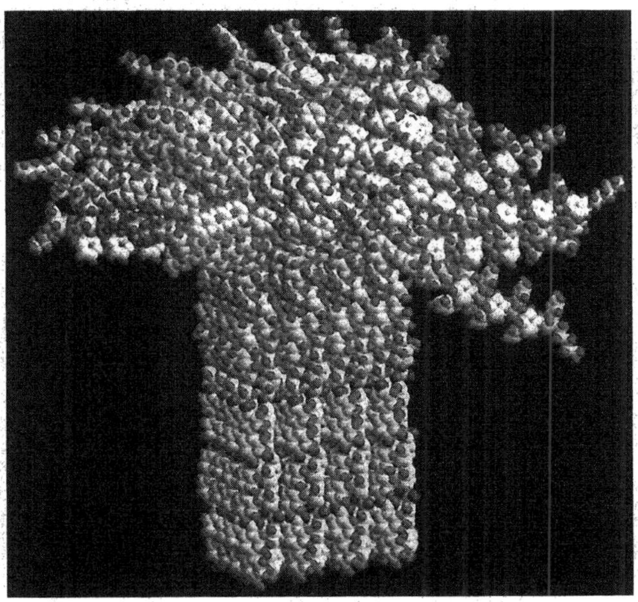

Figure 1.5. Supramolecular nanostructure formed by the ordered self-assembly of triblock copolymers (reprinted with permission from Stupp 1997, ©1997 American Association for the Advancement of Science).

1.7.6 Super Strong Materials by Nanostructuring

Contact persons: H. Kung and T.C. Lowe, Los Alamos National Laboratory

Traditionally, the mechanical strength, σ, of crystalline materials is believed to be largely controlled by the grain size d, often in the manner described by the Hall-Petch relationship, $\sigma = kd^{-1/2} + \sigma_0$. As the structural scale reduces to the nanometer range, researchers have found that the materials exhibit different scale dependence and there is a limit to the conventional descriptions of yielding (Misra et al. 1998). In addition to the high strength, the intrinsically high interface-to-volume ratio of the nanostructured materials may enhance interface-driven processes to extend the strain-to-failure and plasticity. A recent study on nanostructured Cu/Nb composites shows a complete suppression of brittle fracture when the wire was tensily tested at liquid He temperature (Han et al. 1998). This is an amazing finding, since bcc metals (such as Nb) are known to fracture in a brittle fashion at 4.2 K. The nanostructured Cu/Nb composites exhibit significant strain hardening and ductility before fracture at a tensile strength of ~2 GPa and a strain of 10.

These results show that by reducing the structural scale to the nanometer range, one can extend the strength-ductility relationship beyond the current engineering materials limit, which is illustrated by the broad curve in the schematic diagram on the right side of Figure 1.6. A limitation of current engineering materials is that gain in strength is often offset by loss in ductility. The nanocomposite results suggest that by reducing the structural scale and by fully understanding the deformation physics governing the plasticity processes in nanostructured materials, we can produce materials with a combination of high strength and ductility (top right-hand corner of Figure 1.6).

Due to their extremely complex nature and ultrafine structural scale, the characterization of deformation physics of nanostructured materials requires a close integration of state-of-the-art experimentation with atomistic modeling. Three-dimensional molecular dynamic (MD) simulations with up to 100 million atoms can now be performed with realistic interatomic potentials based on the embedded-atom method (Zhou et al. 1998; Holian and Lomdahl

1998). This opens up possibilities for unprecedented direct comparison between theory and experiments that will greatly enhance our understanding of the fundamental physics of materials with strength close to the theoretical limits.

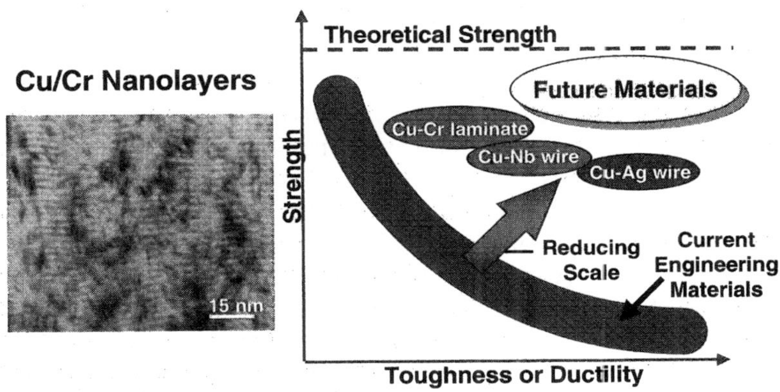

Figure 1.6. High resolution TEM image (left) of Cu/Cr nanolayers and diagram (right) of how reducing the scale of wire (materials) structure will affect the "toughness" and ductility of the materials. Nanostructured materials of the future will be able to transcend the limits of strength and ductility of current engineering materials.

1.7.7 Quantum Computing

Contact person: H. Everitt, U.S. Army Research Office

One of the primary justifications for investment in nanotechnology stems from the desire to continue "Moore's law" which, in one form, states that the feature sizes of microelectronic devices shrink by half every four years. At this rate, feature sizes will be less than 10 nanometers by 2020 and atomic scale by 2035. In fact, Moore's law will halt before then, about 2012, because of quantum mechanical effects that will prevent us from continuing to improve performance of logic devices simply by shrinking them. At that point, new information processing methodologies will be required if we are to continue to advance our ability to compute.

Perhaps the most promising approach to getting beyond Moore's law is that of quantum computing. Originally proposed by Richard Feynman in the early 1980s, the idea is to take advantage of quantum mechanics, rather than be limited by it, to develop processors that simulate physical phenomena more naturally and exponentially faster than a digital computer can. Whereas a digital bit may only store information in the form of a sequence of "0s" and "1s," a quantum bit may be in a superposition state of "0" and "1," that is, representing both values simultaneously until a measurement is made. A sequence of N digital bits can represent a single number between 0 and (2^N)-1, while N quantum bits can represent all 2^N numbers simultaneously. A quantum computer with only 300 quantum bits can represent a system with $2^{300} \sim 10^{100}$ elements, a number greater than the number of atoms in the universe! A quantum computer could solve problems much more complex than a digital computer ever could.

Quantum algorithms have been developed to factor large numbers exponentially faster than digital computers could and to search N-element databases at a rate $N^{1/2}$ faster than a digital computer could. However, to date only simple quantum logic operations on a few, atom-based quantum bits have been demonstrated, and it is clear that this approach will not lead to large scale quantum computation. Nanotechnology may solve the problem of fabricating multiple quantum bits just as it brings us to the end of Moore's law. Arrays of semiconducting or superconducting quantum dots ("artificial atoms") are within reach of today's nanotechnology. Quantum dots can be used for quantum bits if (1) they are nanometer sized to exploit quantum mechanical effects, (2) identical to every other quantum dot in the array, and (3) isolated from the environment to preserve the quantum effects. Demonstrations of single quantum dot quantum bits may only be a few years away; if so, nanotechnology will provide the means to fabricate arrays of thousands of quantum bits, and the first large scale quantum computers may not be far behind.

It has been claimed that a classical computer could take as long as the age of the universe to factor a 200-digit number into its two prime cofactors. This is the basis of modern cryptography: the cofactor is the key; the 200-digit number is broadcast openly with the foreknowledge that the number won't be factored in time to be useful to an eavesdropper. It is possible that a quantum computer could factor that number in minutes, rendering today's most sophisticated cryptographic schemes vulnerable and obsolete.

1.7.8 Three-Dimensional Structures for New Circuitry

Contact person: M. Reed, Yale University

Figure 1.7 depicts a collection of nanoparticles about 5 nm in diameter (gold particles in gold and semiconductor particles in green) with attached functionalized endgroups (colored rods on particles) and buckyballs and nanotubes (black) on a substrate with gold electrodes. This is a rendition of a self-assembled molecular circuit produced by self-assembly of the molecules and the ~5 nm gold nanoparticles using a thiol (S) chemistry process in a beaker.

Figure 1.7. Collection of nanoparticles of about 5 nm dia. (©1999 Mark Reed. All rights reserved).

1.8 REFERENCES

Dunin-Borkowski, R.E., M.R. McCartney, R.B. Frankel, D.A. Bazylinski, M. Posfai, and P.R. Buseck. 1998. Magnetic microstructure of magnetotactic bacteria by electron holography. *Science* 282:1868-1870.

Han, K., J.D. Embury, J.R. Sims, L.J. Campbell, J-J. Schneider-Muntau, V.I. Pantsyrnyi, A. Shikov, A. Nikulin, and A. Vorobieva. 1999 (in press). The fabrication, properties and microstructures of Cu-Ag and Cu-Nb composite conductors. *Materials Sci. and Eng. A.*

Holian, B.L. and P.S. Lomdahl. 1998. Plasticity induced by shock waves in nonequilibrium molecular dynamics simulations. *Science* 280:2085-2088.

Misra, A., M. Verdier, Y.C. Lu, H. Kung, T.E. Mitchell, M. Nastasi, and J.D. Embury. 1998. Structure and mechanical properties of Cu-X (X = Nb, Cr, Ni) nanolayered composites. *Scripta Mater. V* 39:555-560.

Stupp, S.I., V. LeBonheur, K. Walker, L. S. Li, K. E. Huggins, M. Keser, and A. Amstutz 1997. Supramolecular materials: Self-organized nanostructures. *Science* 276:384.

Zhou, S.J., D. Preston, P.S. Lomdahl, and D.M. Beazley. 1998. Large-scale molecular dynamics simulations of dislocation intersection in copper. *Science* 279:1525-1527.

Chapter 2

INVESTIGATIVE TOOLS: THEORY, MODELING, AND SIMULATION

Contact persons: **D. Dixon, Pacific Northwest Laboratory; P. Cummings, University of Tennessee; K. Hess, University of Illinois, Urbana**

2.1 VISION

A critical issue for nanotechnology is that the components, structures, and systems are in a size regime about whose fundamental behavior we have little understanding. The particles are too small for direct measurements, too large to be described by current rigorous first principle theoretical and computational methods, exhibit too many fluctuations to be treated monolithically in time and space, and are too few to be described by a statistical ensemble. Fundamental understanding and highly accurate predictive methods are critical to successful manufacturing of nanostructured materials, devices, and systems.

2.2 CURRENT SCIENTIFIC AND TECHNOLOGICAL ADVANCEMENTS

In the field of theory, modeling, and simulation (TM&S), the most significant advancements applicable to nanotechnology have been associated with the introduction of more powerful computers and corresponding advances in software and algorithms, and to a lesser degree, with the broaching of new theories. These have enabled the merging of several different types of computational techniques (for example, quantum chemical and molecular dynamics) to provide high-fidelity simulations of nanoscale objects based on first principles theory.

2.3 GOALS FOR THE NEXT 5-10 YEARS: BARRIERS AND SOLUTIONS

Even though it is difficult today to fully model nanoscale systems, it is clear that TM&S is the key enabling field for the following:

- Reducing the time needed to design new materials
- Developing nanoscale devices from the new materials
- Increasing the reliability and predictability of the operation of the new devices
- Designing and optimizing new nanoscale technologies

As is well known from conventional silicon technology, TM&S immediately leads to financial savings in terms of time expended, capital invested, and the quality of the final product. TM&S is an enabling field in that it can provide new levels of understanding and provide numerical values when experimental results are not available. Furthermore, modeling and simulation are often needed to properly interpret an experimental measurement, due to the complexity of the measurement or the averaging done under the experimental conditions. In addition, TM&S is essential to continuing to exploit the living cell's nanosystems as models for future developments in nanotechnology. Recall that much of our current inspiration for the design of many nanodevices and systems has been based on our understanding of nature's nanomachines, proteins.

However, TM&S as applied to nanosystems needs significant advances in order to be successful. For example, quantum chemical and molecular theory and simulation will be required to offer fundamental insights and provide predictability methods for nanoscale material properties such as thermophysical, thermochemical, electrical, magnetic, and rheological behavior. Optimization of nanoscale materials or devices will require exploration of thousands of design alternatives prior to synthesis. Properties of nanoscale devices embedded in larger-scale environments need to be modeled; hence models must traverse the various scales (molecular, nano, meso, and macro) that constitute a working device and its manufacture.

Significant technical issues in modeling nanomaterials and nanodevices add to the complexity of the models. One cannot just consider the individual components; nanoscale devices need to be understood in the context of their environment and as elements of architectures that can include meso-scale and macro-scale entities. An example of such a hierarchy for electronics is

Materials → Devices → Circuits → Systems → Architectures

The tools that are the mainstay of the CMOS-based microelectronics industry today, such as SPICE, PISCES, and SUPREM, have no applicability in the coming nanoelectronic era. We need to develop new tools that connect across all of the various scales in order to enable the electronics industry to advance into the nano era. An example of new technology is that the ever-increasing speed of computers has enabled us to find numerical solutions to several differential equations, including Shockley's semiconductor device equations, Maxwell's equations, Boltzmann's equation, and Schroedinger's equation (for given potentials), all with realistic boundary conditions in three dimensions.

Enhanced computing capability thus enables the researcher to simulate electronic devices beyond the conventional way in which the device function has been partitioned into various subfunctions and sections. The physics of these sections was once explored using modest numerical tools such as integration and compared to experiments; the connection to other device sections and creative optimization was entirely left to the researcher. It is now possible to treat the device as a whole, and the interactions between the various device functions, parts, and sections can be optimized numerically, resulting in important improvements that are really needed from an engineering point of view.

For future nanostructure simulation, these new tools and methods must be combined and linked to an atomistic understanding. An interdisciplinary effort will be required to combine solutions of the equations of Shockley and Boltzmann (as done now in electrical engineering) with atomistic density functional theory (as done now in chemistry and physics). Unique opportunities exist here for the merging and novel use of existing knowledge.

There are many length and time scales that are important in nanotechnology. The length scale goes from 10 Å to 10^4 Å, which encompasses 10^2 to 10^{11} particles, yet one must still be faithful to the atomic scale, since nanomaterials will often be made up of small molecules. Because the particles are small, surface effects are crucial. Interactions with other species are also crucial, as are interactions with the environment; this means that researchers must consider using the chemical potential where appropriate. There are also many different types of time scales, ranging from 10^{-15} s to several seconds, so consideration must be given to the fact that the particles are actually undergoing fluctuations in time and to the fact that there are uneven size distributions. To provide reliable results, researchers must also consider the relative accuracy appropriate for the space and time scales that are required; however, the cost of accuracy can be high. The temporal scale goes linearly in the number of particles N, the spatial scale goes as O(NlogN), yet the accuracy scale can go as high as N^7 to N! with a significant prefactor. A critical problem is that there are fundamental limitations on the ability of parallel supercomputers to solve the time problem, because this is dependent on the speed of the processor as well as on the speed of the switch. Furthermore, much of the action on the nanoscale takes place at the transition/interface between quantum mechanics and classical mechanics, a region for which not many methods have been developed and which has not been explored in detail.

2.4 SCIENTIFIC AND TECHNOLOGICAL INFRASTRUCTURE

The TM&S areas that must be developed in order to make needed advances in nanoscience include mesoscale theories; complexity theory; multiscale methods, including advances in applied mathematics; and order-N methods for computational efficiency—all with the required accuracy. In addition, there is a critical need for development of accurate, transferable force fields for molecular simulations for all atoms in the periodic table. The complexity of molecular systems also demands new methods for optimizing complex structures that will have application in predicting the self-assembly of nanostructured materials; an example of this is *de novo* protein folding. Furthermore, all of these methods need to be interoperable, and the data from one calculation must be able to be readily used by another, in the correct form, with the correct accuracy and uncertainties included. For example, the computational chemistry/ electronics/micromechanics/physics (CCEMP) approach needs to be developed beyond its current capabilities, since self-consistent combinations of the relevant differential equation systems need to be solved. As another example, a laser diode demands the solution of the Shockley equations, the Maxwell equations, and the Schroedinger equation for the quantum well, all fully and self-consistently coupled.

Even if such solutions were possible with current computers—and in three dimensions they are not—this still would not solve all nanostructure problems. The various methods must be integrated and made to be easily used by a broad range of scientists and engineers in order to have an impact on device design. Achieving this will necessitate development of modern collaborative problem solving environments (CPSE), as well as shared databases. A CPSE is really needed in order to have a geographically dispersed group of people all working on the same problems and developing a common set of theories and software. We must build smart software that automatically adjusts the level of calculation for the best balance of speed and accuracy; we must build interfaces so that an engineer can design systems on a computer without a detailed understanding of every aspect of quantum mechanics, force fields, molecular dynamics, mesoscale, finite element analysis, etc.; the designers also need answers fast enough to make decisions. An example of a CPSE would be an Internet platform that makes the existing simulation tools (e.g., band-structure calculations, Poisson solvers, etc.) available to everyone. This should include (at least a list of) solvers of Maxwell's, Shockley's (devices), Boltzmann, and Schroedinger equations.

Clearly, most areas of nanotechnology will be dramatically improved by increased emphasis on simulation. However, there can be some misunderstanding of the nature of the critical needs in this area. In general, there is a strong emphasis on computer science, particularly the acquisition of massively parallel hardware and software compatible with massive parallelism. Such supercomputer facilities are essential for the large-scale simulations required to advance the technology and for providing a number of the solutions required.

It is also important to support development of theoretical methods, which are just as important in this area as is developing the software. We must also train students in TM&S for nanotechnology in a wide range of areas, including chemistry, physics, materials science, applied physics, biology, computer science, chemical engineering, and electrical engineering, and provide them with backgrounds in ceramics, polymers, semiconductors, metal alloys, catalysts, and other areas.

2.5 R&D INVESTMENT AND IMPLEMENTATION STRATEGIES

A concerted coordinated interagency basic research effort in nanoscience and technology would have enormous synergy with the presidential interagency initiative Information Technology for the Twenty-First Century (IT^2). Nanoelectronic devices are the *only* apparent route to creating the computing and networking breakthroughs needed to fulfill the ultimate goals of the IT^2 initiative. Likewise, the extraordinary advances envisioned in the IT^2 initiative provide the computer power and software environments that permit computational modeling and simulation on the extreme scale needed to understand, design, and optimize new nanoscale devices. Thus, realization of the goals of the IT^2 initiative is crucially dependent on success in an aggressive national initiative in nanoscale science and technology; likewise, the full potential of a national initiative in nanoscale science and technology cannot be achieved without the development of new computational and simulation capabilities made possible through the high-performance computing and networking resources provided by the IT^2 initiative.

2.6 PRIORITIES AND CONCLUSIONS

Priorities

- Develop simulations that embody multiscale and coupled multiphenomena descriptions. Special attention should be given to larger systems of atoms and molecules and to simultaneous simulation of more than one aspect (mechanical, electronic, etc.).
- Advance outstanding theories, such as nucleation, charging, electron transport, mechanical cracking, chemical reactions in special environments, and multibody processes in order to significantly improve simulations.

- Take advantage of outstanding opportunities for development of theories and simulation methods by studying nanostructures. This may provide ground for new theories that can be used in other fields.
- Maintain continuous assessment and interaction among researchers doing simulations on various aspects of nanoscience and engineering.
- Encourage cross-disciplinary research (e.g., nanoelectronics, thermodynamics, ehcmistry, mechanics, biological processes, and others) and education in order to enable future developments in nanotechnology.

2.7 EXAMPLES OF CURRENT ACHIEVEMENTS AND PARADIGM SHIFTS

2.7.1 Scales and Scaling of Computational Complexity for Molecular Calculations on Nanoparticles

Contact person: D. Dixon, Pacific Northwest Lab.

One of the important issues in computational science is the scaling of the computation. As shown in Figure 2.1(a), scaling must be considered in three different dimensions. Figure 2.1(b) and (c) show details of the different experimental and computational methods that can be used to study the temporal axis (b) and the spatial axis (c) (Alivisatos et al. 1998).

In (a), the X axis represents spatial scaling and ranges from the size of an atom (~1 Å) to the size of a nanoparticle (~1 μm). The computational scaling for chemical calculations along this axis is reflected by the need to treat Coulomb interactions, and this scales between n and n^2, with the best being O(NlogN) where N is the number of particles. With present treatments of dynamics, temporal scaling is given along the Y axis and is linear in the number of particles; however, there is a significant dynamic range because atomic motions are on the order of 1 fsec which means that 15 orders of magnitude are needed in order to reach a time of 1 sec, still fast for many macroscopic processes. The Z axis represents the accuracy of the calculation. There is an increased need for accuracy in the computation as one strives for tighter design principles, providing new insight, and minimizing the number of expensive experiments. It is now possible to go beyond the old limit of "chemical accuracy" of 1 kcal/mol. The scaling on the Z axis is much worse in terms of the scaling with the number of particles. Actually, the scaling is in terms of the number of basis functions, with between 50 and 100 basis functions needed for each particle. The scaling goes as N^m, with m=7 for very high accuracy calculations, and can go to N! for full configuration interaction calculations. Even accuracies of 5 to 10 kcal/mol may require calculations scaling with m=5. The drive for larger computer resources is driven by the need to go to larger sizes, longer times, and higher accuracy.

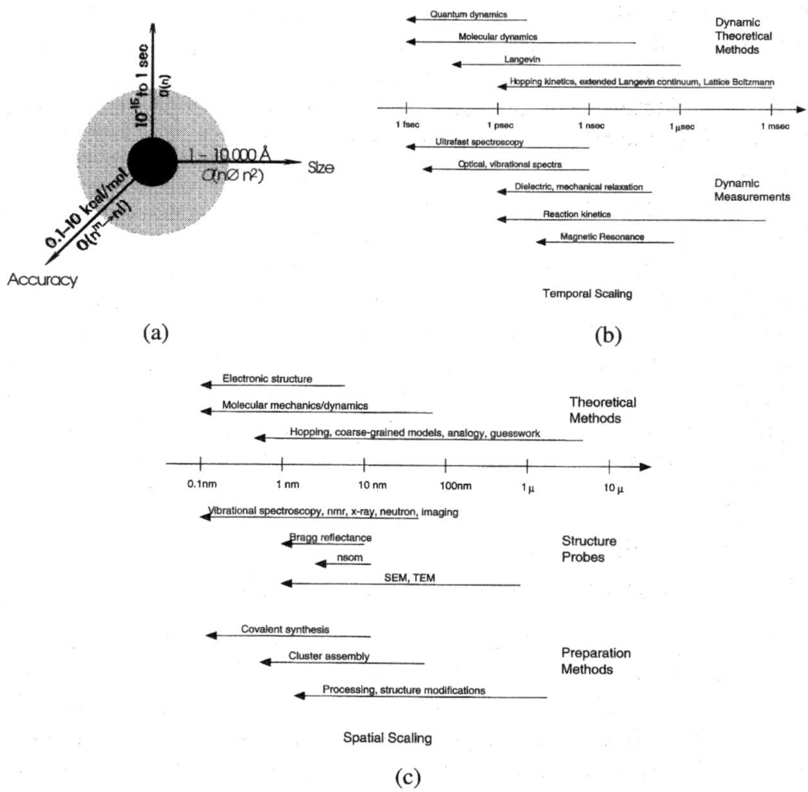

Figure 2.1. Scales and scaling of computational complexity for molecular calculations on nanoparticles.

Although there are classes of problems in nanoparticles that are accessible via established computational and experimental approaches, it is important to emphasize that the questions that become relevant are often not obtainable in terms of well-developed approaches. Figure 2.1 (b) and (c) show why this is so, illustrating the appropriate time and length scales for the understanding of structure and properties of nanoparticles. There are many choices of experimental methods for the preparation and characterization of particles dependent on the length and time scale, and the same is true for theoretical/computational studies. Figure 2.1 (b) characterizes the different temporal regimes that can be studied by using dynamics theory.

Whereas molecular dynamics simulations can be successfully used to examine local, relatively short-time-scale motions, this approach is limited on the full scale of relevant motions. Since relaxation times in condensed phase materials such as polymeric systems extend from picoseconds to hours, many of the important questions are simply not susceptible to classical molecular dynamics. As an alternative, one can turn to a coarser scale, considering Brownian-like motion such as described by Langevin dynamics. It is clear that progress in the area of nanoparticles demands that multiple theoretical approaches be brought to bear on single problems, and that diverse approaches be encouraged. At the same time, there remain opportunities for advancing relatively established methods to improve access to larger systems and longer times. Examples include the development and implementation of methods with linear size scaling in quantum chemistry and methods with multiple time steps and approximations for evaluation of long-range forces in molecular dynamics.

2.7.2 Nanoscale Lubrication

Contact person: P. Cummings, University of Tennessee

A common feature of nanoscale mechanical devices is their need for lubrication between moving surfaces separated by nanometer-sized gaps. For example, the operation of a magnetic recording device relies on the relative motion of the magnetic head and the recording media with a spacing on the scale of nanometers, producing shear in the fluid between the surfaces. The need to achieve higher linear recording density and thus high data storage is pushing this spacing to less than 25 nm, typically using a non-Newtonian liquid-bearing interface to achieve low friction during high speed operation. Common liquid lubricants are perfluoropolyethers, whose steady-state shear rate is as high as ~10^8 s^{-1}, with extremes of up to two orders of magnitude higher, well beyond the range of strain rates accessible to current experimental characterization methods. Similar shear rates arise in the lubrication of many microelectromechanical systems (MEMS), including micromotors.

In recent years, surface force apparatus (SFA) experiments on model lubricants, including alkanes, have provided intriguing insight into nanoscale lubrication, suggesting that the viscosity of confined nanoscale lubricant films is orders of magnitude higher than that of the same bulk lubricant, and that the nano-confined lubricants are non-Newtonian (shear-thinning) over ranges of strain rate (10^2-10^5 s^{-1}) for which the corresponding bulk lubricant is Newtonian. This has raised questions about the feasibility of lubricating MEMS devices and concerns that high power consumption (relative to size) will be required to overcome friction at start-up and viscous resistances during operation. Careful molecular dynamics simulations at high strain rates (~10^8 s^{-1} and higher) have shown little difference between confined and bulk behavior (Figure 2.2 and Figure 2.3).

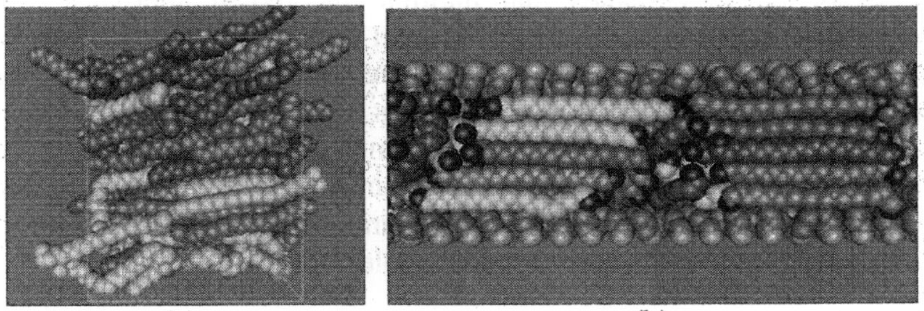

(a) (b)

Figure 2.2. (a) Snapshot of n-tetracosane in bulk under shear flow at a shear rate of 7 x 10^9 s^{-1}, density of 0.82 g cm^{-3} and temperature of 313 K. Only the carbons are shown. Molecules are shaded in different colors to aid in distinguishing them. (b) Snapshot of n-tetracosane under shear flow confined between two walls with tethered butane chains, with wall spacing of 3.6 nm, the same apparent shear rate, core liquid density of 0.82 g cm^{-3}, and temperature of 313 K. End groups are shown in black.

Figure 2.3. Snapshot of squalane under shear flow confined between two walls with tethered butane chains, with wall spacing of 3.6 nm, the same apparent shear rate as in Figure 2.2, core liquid density of 0.82 g cm^{-3}, and temperature of 323 K. Sidegroups are shown in gray.

However, there is a fundamental disconnect between the strain rates accessible to molecular simulation (~10^8 s^{-1} and higher) and to experiment (~10^5 s^{-1} and lower). The disconnect can only be bridged by much larger scale molecular dynamics calculations on future massively parallel supercomputers permitting accurate simulations at lower strain rates and extensions of the SFA experiments to higher strain rates. A similar problem occurs on simulations of glasses in terms of reaching the glassification temperature.

2.7.3 Simulations of Carbon Nanotubes

Contact person: M. Meyyappan, NASA Ames

Carbon nanotubes (CNT) exhibit extraordinary electronic and mechanical properties. In the past few years, simulations have been valuable to compute the properties of CNT, evaluate potential applications, and design interconnects, sensors, and functional devices. The electronic structure of CNT can be either metallic or semiconducting depending on both the diameter and chirality of the tube. The possibility of connecting nanotubes of different diameter and chirality has the potential to create heterojunctions leading to functional electronic devices, logic gates and circuits. Figure 2.4 (Menon and Srivastava 1997) shows formation of a metal-semiconductor-metal T junction using (5, 5), (10, 0), and (5, 5) nanotubes. This structure was optimized using a generalized tight-binding molecular dynamics scheme.

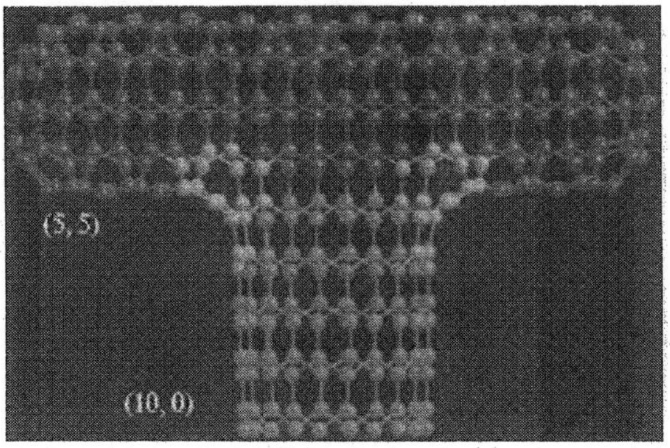

Figure 2.4. Metal-semiconductor-metal T junction using (5, 5), (10, 0), and (5, 5) nanotubes. The turquoise colored balls denote the atoms forming the heptagons (Menon and Srivastava 1997).

Molecular machines in the simplest form of molecular bearings, shaft and gear, and multiple gear systems are of interest since they can lead to functional nanoelectromechanical systems (NEMS). Figure 2.5 shows a carbon nanotube-based gear, which is 2 nm in diameter. Shafts are single-walled carbon nanotubes and gear teeth are benzyne molecules bonded onto the nanotube. This is a simple, synthetically accessible structure. Molecular dynamics simulations were used to investigate the properties and design space of the CNT gears.

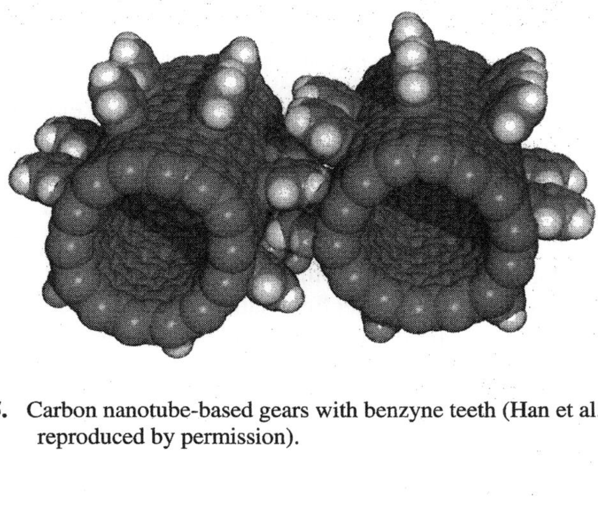

Figure 2.5. Carbon nanotube-based gears with benzyne teeth (Han et al. 1997; reproduced by permission).

2.7.4 Simulation of Quantum Dots

Contact person: K. Hess, University of Illinois, Urbana

The quantum dot will be a basic element of future devices. The quantum dot can be considered as a large atom. For a small number of electrons, one can use quantum Monte Carlo methods and for larger systems, one can use some form of electronic structure theory, for example, density functional theory (DFT). Figure 2.6 shows an atomistic SiO_2 quantum dot bridging between two silicon surfaces. It is meant to symbolize the connection of nanoelectronics to conventional (silicon—silicon dioxide—silicon) electronics as well as the multiscale nature of the problems (quantum dot connected to bulk). Calculations on this system have been carried out at the DFT level.

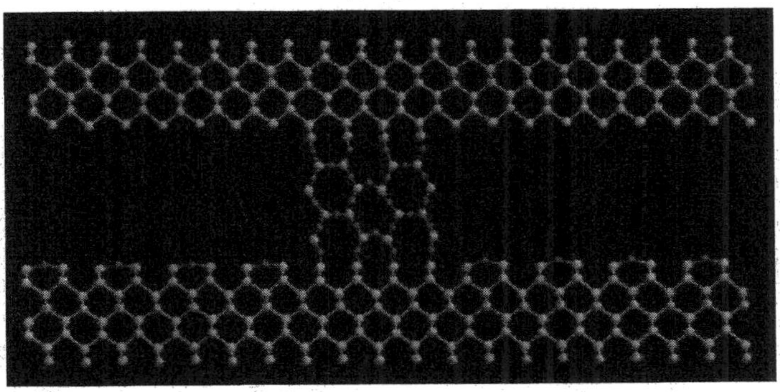

Figure 2.6. An atomistic SiO_2 quantum dot bridging between two silicon surfaces.

2.7.5 Molecular Simulation of DNA Molecule Dynamics

Contact person: Tamar Schlick, New York University

The study of DNA supercoiling—the large-scale folding of the DNA double helix upon itself—is of wide interest because supercoiled DNA is a key functional form of DNA in the cell, directing and facilitating many fundamental biological processes such as replication, recombination, and transcription (Schlick 1995). The problems involved also have practical significance because an improved understanding of the topological and geometric aspects of DNA supercoiling can impact the design of vectors for gene therapy and of topoisomerase inhibitors that act as anticancer or antibacterial drugs.

Kinetic aspects of DNA supercoiling are difficult to measure experimentally on short timescales, and thus simulations can shed insights into the processes and mechanisms involved. The influence of superhelicity on the specific problem of site juxtaposition was studied using a macroscopic wormlike chain/bead model of DNA that considers thousands of base pairs (the DNA diameter is around 2 nm and the length when stretched is 340 nm per 1,000 base pairs) in combination with an efficient Brownian dynamics algorithm. The potential energy accounts for bending and torsional deformations, screened electrostatic terms, and hydrodynamic interactions. This combination makes possible simulations over millisecond time frames that account for key factors that affect DNA conformational flexibility. Previously impossible measurements can be made regarding time estimates for site juxtaposition. Site juxtaposition is the process of bringing linearly distant DNA segments together in space due to the internal dynamics of the DNA molecule. Based on analyses of trajectories such as shown in Figure 2.7 for 3,000 base pairs, the site juxtaposition times can be determined as a function of the superhelicity, the salt concentration, and the separation distance between the sites. These measurements help in the analysis of site-specific recombination reactions to provide lower time bounds for site synapsis (Jian et al. 1998).

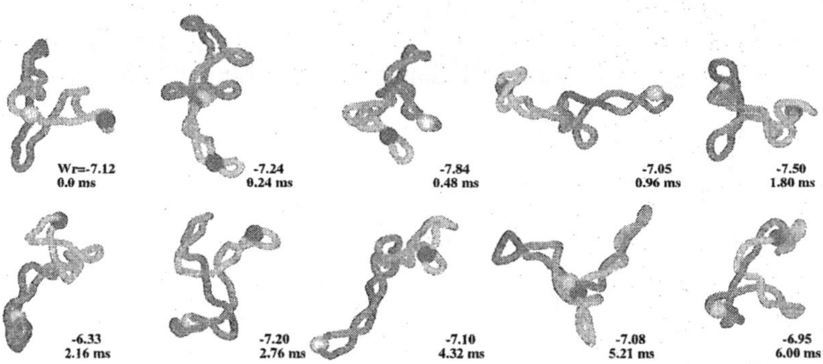

Figure 2.7. Brownian dynamics snapshots of DNA: Brownian dynamics snapshots of 3,000 base pairs of circular DNA focusing on the juxtaposition of two segments separated by 1,200 base pairs over 6 ms, showing the large random motions that bring sites together (figure prepared by Jing Huang).

Mechanistic details of the dynamic process could also be inferred. Current studies involve extensions to DNA systems of size 12,000 base pairs, and planned work involves incorporation of DNA sequence effects.

2.7.6 Simulation of quantum confinement in silicon nanocrystals

Contact person: J. Chelikowsky, University of Minnesota

Naturally occurring silicon has a low radiative efficiency in the optical region. For this reason, although silicon is the material of choice for making electronic devices, it is not a good material for making optical devices such as solar cells and lasers. Its optical gap is too small. This is in contrast to other semiconductors like gallium arsenide, which has a larger optical gap. As a consequence, gallium arsenide is a material of choice for lasers and other optical applications. If the properties of silicon could be altered to resemble gallium arsenide, then one would have a universal material for building all opto-electronic devices. This is an important consideration, as a huge investment has been made in processing and manufacturing silicon-based devices.

Recently, it has been discovered that the optical properties of silicon can be altered by confining the optical excitation to a small region of space, e.g., within a large cluster of silicon atoms. These large clusters are called quantum dots. The dots are fragments of crystalline silicon that have been terminated by a passivating material such as atomic hydrogen. Such a quantum dot is illustrated in Figure 2.8. Its optical gap measured as function of dot size is illustrated in Figure 2.9. The optical excitation for the quantum dot is significantly enhanced, both in frequency and in intensity. This effect, called quantum confinement, makes silicon resemble gallium arsenide in terms of its optical properties. For example, the excitation energy for the silicon dots in Figure 2.9 is as large as 2-2.5 eV, twice the optical gap of crystalline silicon, and comparable to the gap of 1.5 eV in gallium arsenide. First principles, parameter-free calculations for the optical gap in silicon quantum dots have been performed to verify the role of quantum confinement and to assess the experimental interpretation. Using massively parallel computer platforms and newly developed algorithms to examine such large systems, clusters up to 30 Å in diameter can be handled.

The largest cluster considered was $Si_{575}H_{276}$ (Figure 2.8). This dot contains several thousand electrons; consequently, the change in energy upon absorption of light is very small relative to the total energy of the electrons. By focusing only on the role of the electronically active states, this calculation becomes tractable. This is the first time optical excitations have been computed for a dot of this size. In Figure 2.10, the calculated optical gaps are compared to experiment. The overall agreement is quite good. This work has led to a better understanding of the role of quantum confinement and a means of predicting optical transitions in nanostructured matter.

Figure 2.8. Ball and stick model of $Si_{575}H_{276}$. The white dots correspond to H atoms, which passivate the surface of the dot.

2.7.7 Molecular Dynamics Simulation of Piezoelectric Polymers

Contact person: Jeffrey Hinkley, NASA Langley Research Center

Researchers at NASA Langley Research Center (LARC) have simulated the molecular motion during poling of an amorphous polyimide. The approach involved parameterizing interatomic forces and atomic charges using quantum mechanics, verifying against experimental measurements of density and dipole moment, and predicting dipolar response to an electric field and the dielectric relaxation strength. Based on insights from this simulation, dozens of proposed new polymers have been rapidly screened, saving considerable effort in the laboratory. Similar methods can be applied to calculations of solubility, interactions with surfaces (adhesion), and permeability to gases such as hydrogen propellant.

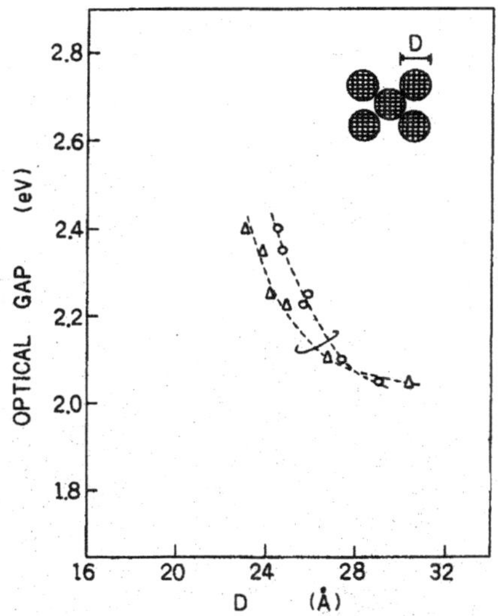

Figure 2.9. Experimental optical gaps for silicon quantum dots. D is the diameter of the dot. The results are from Furukawa and Miyasoto (1988). The two sets of experimental data represent different estimates for the size of the dots.

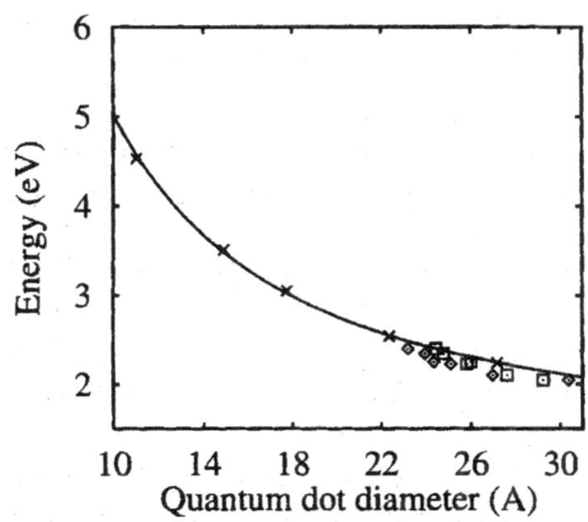

Figure 2.10. Theoretical optical gaps versus experiment. The theory (solid line) is from Ogut, Chelikowsky, and Louie (1997). The open symbols are from experiment; see Figure 2.9.

The simulation shown in Figure 2.11 is used for lightweight sensors for integrated smart materials. In this case, the breakthrough is observation of piezo activity up to 100°C above temperature limits of state-of-the-art materials. Future directions include simulation of semicrystalline materials, nanocomposites containing layered silicates, and revolutionary organic/inorganic hybrids.

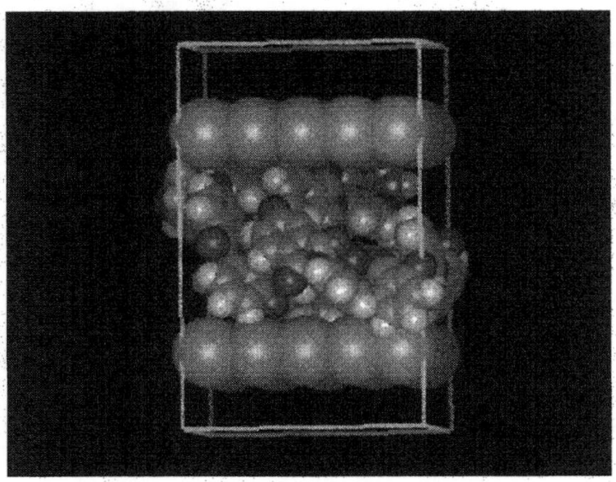

Figure 2.11. An amorphous cell at experimental density containing 5 repeat units of a piezoelectric polyimide. Planes of large spheres represent metal electrodes; they are used to simulate a poling field.

2.8 REFERENCES

Alivisatos, A.P., P.F. Barbara, A.W. Castleman, J. Chang, D.A. Dixon, M.L. Klein, G.L. McLenson, J.S. Miller, M.A. Ratner, P.J. Rossky, S.I. Stupp, and M.E. Thompson. 1998. From molecules to materials: Current trends and future directions. *Angew. Chemie, Adv. Mat.* 10:1297-1336.

Furukawa, S. and T. Miyasoto. 1988. Quantum size effects on the optical band gap of microcrystalline Si:H. *Phys. Rev. B* 38:5726.

Han, J., A. Globus, R. Jaffe, and G. Deardorff. 1997. Molecular dynamics simulation of carbon nanotube based gears. *Nanotechnology* 8:95-102. Bristol, U.K.: IOP Publishing Ltd.

Jian, H., T. Schlick, and A. Vologodskii. 1998. Internal motion of supercoiled DNA: Brownian dynamics simulations of site juxtaposition. *J. Mol. Biol.* 284:287-296.

Menon, M., and D. Srivastava. 1997. Carbon nanotube "T junctions": Nanoscale metal-semiconductor-metal contact devices. *Phys. Rev. Lett.* 79 (22):4453-4456.

Ogut, S., J.R. Chelikowsky, and S.G. Louie. 1997. Quantum confinement and optical gaps in Si nanocrystals. *Phys. Rev. Lett.* 79:1770.

Schlick, T. 1995. Modeling superhelical DNA: Recent analytical and dynamic approaches. *Curr. Opin. Struct. Biol.* 5:245-262.

Young, J.A., B.L. Farmer, and J.A. Hinkley. 1999. Molecular modeling of the poling of piezoelectric polyimides. *Polymer* 40(10):2787.

Chapter 3

INVESTIGATIVE TOOLS: EXPERIMENTAL METHODS AND PROBES

Contact persons: J. Murday, NRL; R. Celotta, NIST; D.Y. Pui, U. Minnesota; P. West, ThermoMicroscopes, Inc.

3.1 VISION

The 1993 NSF panel report *Atomic Imaging and Manipulation (AIM) for Advanced Materials* (NSF 93-73) concluded that (a) important new science would become accessible as a result of the development of atomic-resolution microscopy, (b) a substantial program in electron microscopy and scanning tip techniques would strengthen U.S. competitiveness, and (c) many user-friendly, low-cost, fast-turnaround compact microscopes were important for rapid progress in much of materials science (Cohen 1993). These conclusions remain valid, but the range of instruments and measurable properties has been extended. Continued development of new tools is critical to the pace of further progress in nanoscience and technology—they provide the "eyes" to see and the "fingers" to manipulate nanostructures. In the nearer term, the greater need is to provide laboratory researchers with the instruments and tools to discover and investigate new chemical, physical, and biological phenomena and applications. In the longer term, those tools will evolve into inexpensive, easy-to-use sensors and/or diagnostic devices with broad applications.

3.2 CURRENT SCIENTIFIC AND TECHNOLOGICAL ADVANCEMENTS

The recent rapid advances in nanotechnology are due in large part to our newly acquired ability to measure and manipulate individual structures on the nanoscale. Whether it be scanning probes, optical tweezers, high-resolution electron microscopes, or other new tools, instruments available to research workers in science and technology now permit them to create new structures, measure new phenomena, and explore new applications. There are limitations for various properties, such as the chemical composition of a single nanostructure and local electronic and thermal characteristics.

Focused Beams

- *Electrons.* Electron microscopy, long the workhorse of science on the sub-micron length scale, is now capable of imaging individual atoms in nanostructures with sub-angstrom resolution (Cowley and Liu 1993). Elemental information is available from electron energy-loss, Auger and X-ray emission measurements with near atomic resolution (Edgerton 1996). New electron based methods have been used to make significant advances in our understanding of magnetic nanostructures (Mankos et al. 1996).
- *Ions.* Ion beams are available with 10 nm resolution and offer some limited analytical capability (Kalbitzer et al. 1993).
- *Photons.* Visible photons are limited by diffraction to spot sizes much larger than a nanometer, unless one operates in the near field (see Scanning Probe section below). X-ray beams might be focused into nanometer spots. Present technology is closer to 1 micron. The limitations are optical elements effective at the X-ray wavelengths and adequate photon fluxes. Rotating anode X-ray sources can provide bright line radiation, but synchrotron radiation is necessary for variable frequency photons. Focusing can be enhanced through capillary X-ray waveguiding (Yamamoto 1997) or, potentially, by the development of nanostructured optical elements.

Electron Microscopy

Rather than focusing an incident beam, electron optics can be utilized to form high-resolution images with the electrons emitted from a surface (Bauer 1990). Image resolution of 12 nm has been reported for photoemission (PEEM) (Ade et al. 1999). Used in conjunction with the new synchrotron X-ray sources, this allows the imaging of nanoscale features with elemental specificity. A variant, which has important applications in the study of magnetic nanostructures, is X-ray magnetic circular dichroism (XMCD) (Stöhr et al. 1993).

Spectroscopic Scanning Probe Microscopes

The inventions of the scanning tunneling microscope (STM) (Binnig et al. 1982) and the atomic force microscope (AFM) (Binnig et al. 1986) have spawned development of a variety of new scanning probe microscopes (SPMs) (Wickramasinghe 1989; Wiesendanger 1994). As a class, the SPMs measure local properties with nanometer-scale spatial resolution by bringing a sharp tip in proximity (1-10 Å) to a solid surface. The proximity of tip and surface enables the SPMs to operate in ambients forbidden to traditional vacuum-based surface analytical techniques. The STM and the AFM were initially limited to monitoring fine scale topography. But the broader class of scanning probes, derived from these initial instruments, allows one to go beyond topography and examine many other local properties, including the following:

- *Electronic structure* by scanning tunneling spectroscopy (STS) (Stroscio and Kaiser 1993), particularly at low temperatures (Bürgi et al 1998; Yazdani et al. 1997).
- *Optical properties* by near-field scanning optical microscopes (NSOM) (Betzig et al. 1991). The NSOM beats the diffraction limit and allows optical access to sub-wavelength scales (50-100 nm) for elastic and inelastic optical scattering measurements (see Figure 3.1), as well as for optical lithography.
- *Temperature* by scanning thermal microscope (SThM) (Majumdar et al. 1993). The SThM uses a temperature-sensing tip (Figure 3.2) to map temperature fields of electronic/optoelectronic nanodevices (Figure 3.3) and to measure thermophysical properties of nanostructures.
- *Dielectric constants* by scanning capacitance microscopes (SCM) (Williams et al. 1989). Since the capacitance of a semiconductor depends on carrier concentration, the SCM enables the researcher to map out dopant profiles in semiconductor devices with nanometer-scale spatial resolution.

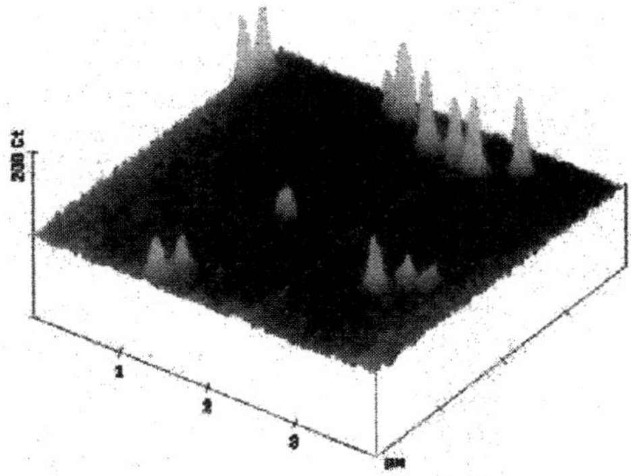

Figure 3.1. Room temperature near-field fluorescence image (4 microns x 4 microns) of single sulfohodamine 101 molecules adsorbed on a silicate glass surface. Each peak is full width half maximum (FWHM) of 100 nm and corresponds to the signal from a single molecule (Bian et al. 1995).

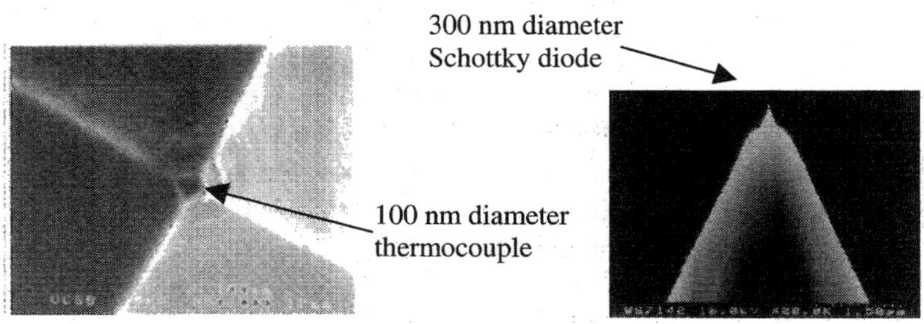

Figure 3.2. Nanofabricated thermocouple (L., reprinted by permission from Luo et al. 1997a; ©1997 American Vacuum Society) and Schottky diode sensors on probe tips (R., Leinhos et al. 1998).

- *Magnetism* by magnetic force and resonance microscopes (MFM) (Hobbs et al. 1989; Rugar et al. 1992). The MFM can image magnetic domains and is already an integral part of characterizing magnetic storage media. The magnetic resonance microscope can detect nuclear and electron spin resonance with submicron spatial resolution and potentially provides a basis for chemical analysis.
- *Charge transfer and the Helmholtz layer* by scanning electrochemical microscope (SECM) (Bard et al. 1991).

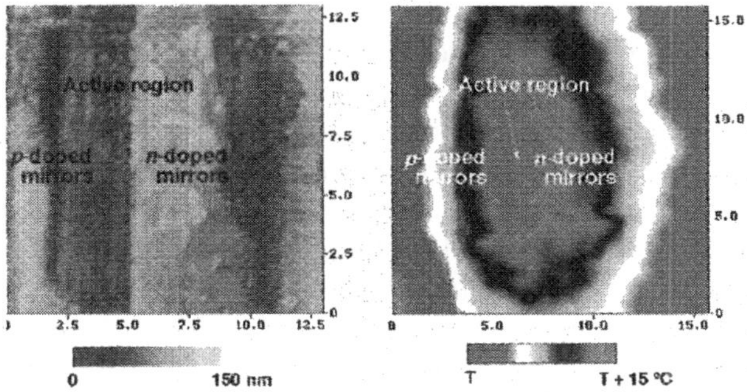

Topography Thermal image

Figure 3.3. Topographical and thermal image of the cross-section of an active vertical cavity laser (reprinted by permission from Luo et al. 1997b; ©1997 American Institute of Physics).

- *Biological molecule folding/recognition* by nanomechanics (Gaub et al. 1997; Colton et al. 1994). Single molecule nanomechanics measurements can provide insights into the molecular phenomena that dominate biological systems and have previously been probed only by measurement of ensemble averages.
- *Chemical information* (Ho et al. 1999; Gimzewski and Joachim 1999; Noy et al. 1997; Knoll and Keilmann 1999).

By providing access to and enabling observation of physical, chemical, and biological phenomena at nanometer scales, SPMs have changed the landscape of experimental research in nanoscience and technology.

Manipulation of Two- and Three-dimensional Nanostructures

Items as small as single atoms and molecules can be manipulated and even exploited as atomic switches (Eigler et al. 1991; Wada 1997). It is interesting to note that atomic manipulation is the smallest possible scale for materials manipulation; we are at a fundamental limit for improving materials behavior through controlling composition and/or structure. There have been many important advances at nanoscale manipulation:

- *Computer-controlled SPM* enables real-time, hands-on human interaction of nanostructure manipulation. In one example, a nanoManipulator (nM) system (Taylor et al. 1993) provides a virtual-environment interface to SPMs; it gives the scientist virtual telepresence on the surface, scaled by a factor of about a million to one. The introduction of direct human-SPM interaction creates not only enhanced measurement capability (for instance, special transducers can provide a sense of touch to the nanomanipulator), but also an automated technology presaging nanofabrication and/or repair of nanostructures. As a demonstration of the educational potential, students in a high school advanced placement biology course have used the nanomanipulator across the Internet to see, feel, and modify Adeno virus particles.

- *Optical tweezers* (Sato and Inaba 1996; Mehta et al. 1999; Kellermayer et al. 1997) provide another new approach to gripping and moving nanometer structures about in three dimensions. This capability has been especially useful investigating particle/ molecular dynamics. A general goal in molecular biophysics is to characterize mechanistically the behavior of single molecules. Whereas past experiments required model-dependent inferences from ensemble measurements, optical tweezers allow a direct observation of the parameters that are relevant to answering the questions, how does a polymer move, generate force, respond to applied force, and unfold?
- *Nanomanipulators* have been reported for use in scanning electron microscopes (SEM) and transmission electron microscopes (TEM). Schmid et al. (1995) have incorporated a manipulating tip, which has 3 degrees of freedom and is controlled to high precision by piezo elements, into a low-energy electron point source microscope. A piezo-driven TEM specimen holder has been made for observing (with atomic resolution) the mechanical interaction in nanometer-sized crystallites, and the mechanical loading and bending of carbon nanotubes (Kizuka et al. 1998; Poncharal et al. 1999). Using this type of specimen holder, the quantized conductance through individual rows of suspended gold atoms has been observed (Ohnishi et al. 1998). A consequence of combining such levels of manipulation with TEM imaging is that the authors were sure (because they directly imaged the Au atom bridges in TEM) of the number of Au atoms in the particular bridge for which they determined the conductance. Still newer, high performance "nano-manipulators" for SEM and TEM (Yu et

al. 1999) have recently been built (see Figure 3.4 and Section 3.7.2 below).

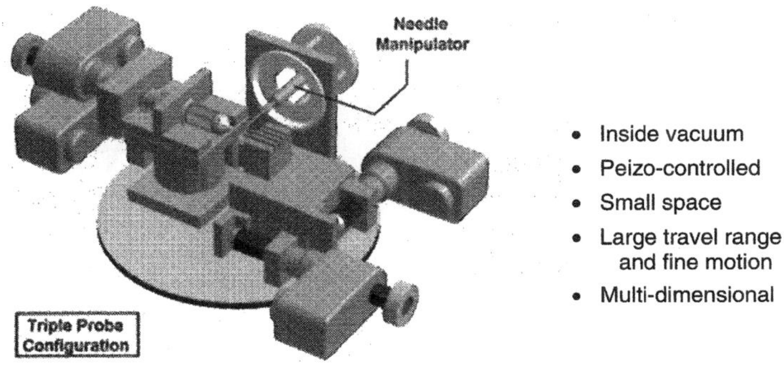

Figure 3.4. Nanomanipulator inside SEM, co-developed by Zyvex and the Rod Ruoff group (see also Yu et al. 1999; reproduced by permission).

Parallel Probe Arrays

Although SPM has been used widely for topographical imaging, atomic/molecular manipulation, and nanoscale lithography, a major drawback is its low raster speed, limited by present cantilever and system dynamics to about 50 Hz/line. To alleviate this problem, several groups are developing arrays of cantilever probes (Figure 3.5) that are individually actuated and controlled (Miller et al. 1997; Minne et al. 1998; Despont et al. 1999). By paralleling the process, they can achieve high-speed nanometer-scale imaging, as well as sub-0.1 μm lithography, on large-scale (1 cm) objects.

In addition to their promise in characterization and fabrication, microfabricated cantilever arrays also show commercial promise as highly sensitive detectors of chemical species (Baselt et al. 1996; Lang et al. 1998).

In-Situ Monitoring and Process Control

Advances in materials processing and fabrication techniques have made it possible to produce superlattice device structures with characteristic layer thicknesses down to several atomic layers and layer interfaces of near atomic precision. Continued demands for improved device performance with simultaneous reduction in production costs have made reproducibility and reliability of superlattice growth vital imperatives. Interfaces must be controlled to atomic dimensions. In situ sensing and feedback control of growth

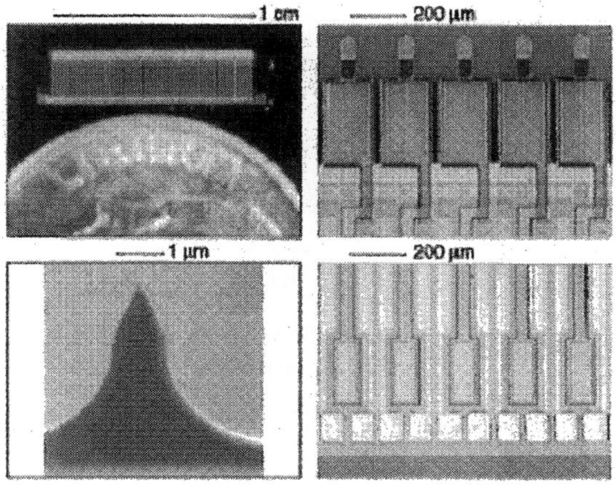

Figure 3.5. An array of cantilevers with integrated actuators and sensors with improved shielding between the actuator and sensor: (a) shows an entire array of 50 cantilevers spanning 1 cm next to a dime; (b) shows a detail of five cantilevers, spaced by 200 μm; (c) is an SEM of a typical integrated single-crystal silicon tip (radius of curvature is below 10 nm); and (d) shows the corresponding electrical contact structure for the cantilevers. There are three leads per device: piezoresistor, ZnO, and tip bias (reprinted by permission from Minne et al. 1998; ©1998 American Institute of Physics).

processes like molecular beam epitaxy (MBE) are essential to reduce the incidence of "nanostructural defects" between adjacent layers. Examples of successful in-situ approaches for monitoring deposition of extremely thin layers include reflection high energy electron diffraction, reflectance spectroscopy, in-situ cathodoluminescence, optical flux monitoring, spectroscopic ellipsometry, photo emission oscillations, absorption band edge spectroscopy, desorption mass spectrometry, vacuum ultraviolet photoionization time-of-flight mass spectrometry, pyrometric interferometry, ultraviolet laser-induced atomic fluorescence, and nonlinear (second harmonic) optical spectroscopies (Schroder 1998).

Nanostructured Materials Characterization

The measurement of nanostructured materials properties is complicated by the presence of aggregated nanostructures. Individual nanoparticles are inherently small, and their compositions and structures are affected by the large number of surface atoms. The particles can be collected into varying degrees of compaction with length scales reaching microns and above (Birringer 1994). Porosity, surface areas, and grain boundaries are susceptible to phase segregation and impurities (Tomkiewicz 1996). Specialized techniques are positron annihilation spectroscopy, small angle neutron scattering, small angle X-ray scattering, wide angle X-ray scattering, extended X-ray absorption fine structure (EXAFS), high resolution electron microscopy, and scanning probes (Edelstein and Cammarata 1996).

3.3 GOALS FOR THE NEXT 5-10 YEARS: BARRIERS AND SOLUTIONS

In general, the goal is development of low-cost, high-resolution, standardized, efficient tools and instruments for manipulation and analysis of nanostructures on surfaces (in two dimensions) and in three dimensions. The following topics are goals for the next 5-10 years:

- *Instruments for analysis of supramolecules, biomolecules, and polymers.* Miniaturized instruments for the analysis of individual molecule properties will be an area of intense research that will have major impact on health, environment, and national security. Microfabricated chips for DNA analysis (Lemieux, et al. 1998; Kurian et al. 1999) and polymerase chain reactions (Kopp et al. 1998) have already been demonstrated. These are the initial steps towards a full-fledged technology of biomedical microdevices, which will not only study and analyze nucleic acids but also other biological molecules such as proteins and carbohydrates. Chip-based sensing for rapid detection of biological pathogens is a critical area with applications in the food handling/processing industry, biological/chemical warfare, and in early warning for exposure to air- and water-borne bacteria, viruses, and other antigens. In one fledgling example, GMR memory elements are being explored for use as biological array detectors (Baselt et al.

1998). Devices such as these require the integration of biology, biochemistry, and surface science with engineering. It is envisioned that biomedical microdevices will be sufficiently inexpensive to make them readily accessible to a large segment of the population, and commonplace in daily life.

- *3-D structure determination.* Present SPMs are limited to analyzing surface or near-surface properties of solids with nanometer-scale spatial resolution. With the exception of the limited capability in ballistic electron emission microscopy (BEEM) (Bell and Kaiser 1996), sub-surface imaging and truly three-dimensional microscopy with nanometer-scale spatial resolution are not currently available; they are, however, extremely important for future development in nanotechnology. For example, most biological nanostructures are three-dimensional and currently imaged by X-ray crystallography, which is expensive and time-consuming. Even in micro/nanoelectronics, which is progressing towards multilayer three-dimensional structures, 3-D imaging would be very useful. Possible approaches for subsurface imaging include ultrasonic echo imaging, non-linear (multiphoton) optical microscopy, and thermal spectroscopic imaging. It is unclear at present which, if any, technique would be suitable. This area clearly needs emphasis.
- *Nanostructure chemical identification.* Chemical identification of an unknown material is crucial to understanding and predicting its properties. Urgently needed are analogs or alternatives to traditional analytical chemistry techniques—elemental analysis (atomic emission spectroscopy, Auger, XPS); mass spectrometry (MS); vibrational (IR, Raman, HREELS); electronic (UV/VIS, UPS); magnetic resonance (NMR, NQR, EPR)—that will work routinely on individual nanometer-sized structures.
- *Functional parallel probe arrays.* Fabrication of probes designed to measure one property has been amply demonstrated; however, full characterization of a nanostructure requires measurement of many properties. One future goal is achievement of multifunctional probes that provide a "laboratory on a tip," or "nanoscale total analysis." Again, this will require integration of knowledge from engineering, chemistry, physics, and biology. As discussed earlier, it is possible to increase the speed of SPMs by

making parallel, individually actuated and controlled probe arrays. Integration of both multifunction and array technologies is likely necessary for realization of rapid nanoscale diagnostics.

- *Standardization and metrology.*
  - Locating and maintaining a position with nanometer accuracy and precision are still difficult, especially if coupled with the requirement to achieve those goals across samples of centimeter dimensions. This is one of the crucial issues that must be solved if commercial nanoelectronic device fabrication is to be realized.
  - Uniform-size nanoparticles of known size and composition are needed for the standardization and calibration of nanoscale measuring instruments.
  - The importance of making measurements on a common set of calibration particles in order to develop reliable standards was underscored by specialists in particles in gases, particles in liquids, particles on surfaces, and mass spectroscopy who came together during a recent DOE Workshop on Instrumentation for Nanoparticles (U. of Minnesota, Dec. 1998). Goals in the next 5-10 years include development of particle-size calibration standards of 3 nm, 10 nm, and 30 nm sizes; improvements in measurement methods for nanometer-size particles, including modeling of the instrument, uncertainty assessment, and improved data analysis methodology; and quantification of uncertainty in TEMs, differential mobility analyzers, and small angle X-ray equipment for measurements over these size ranges. A real-time size distribution analyzer for nanoparticles is needed as a process monitor during processing. Additionally, there is a need for a particle classifier to select nanoparticles into narrow size fractions (Chen et al. 1998).
  - There is an opportunity to define fundamental standards based on the creation of atomically controlled and measured structures (see schematic in Figure 3.6). Quantized electron devices may provide known electrical currents. Macromolecules/clusters of known mass (having a countable number of elemental constituents) may provide building blocks of a gram.

- *New Nano-Manipulators.* Nanotechnology has a goal of 3-D manipulation of chemical moieties to build molecules/clusters and then to assemble them into larger devices and materials. Achieving this requires combining techniques of chemical synthesis with engineering methods that wield atomically precise positional control. Manufacturing technologies such as microelectromechanical systems (MEMS) are potentially capable of producing higher degree-of-freedom micromachines that can exert molecular-level positional control and bridge mesoscopic extremes in handling nanoscale and microscale components. Extension of MEMS into nanometer-sized electromechanical structures (NEMS) will achieve that capability. In combination with chemical functionalization schemes and self-assembly concepts, MEMS/NEMS will form an essential generation of hybrid machines for subsequent stages of nanotechnology development.

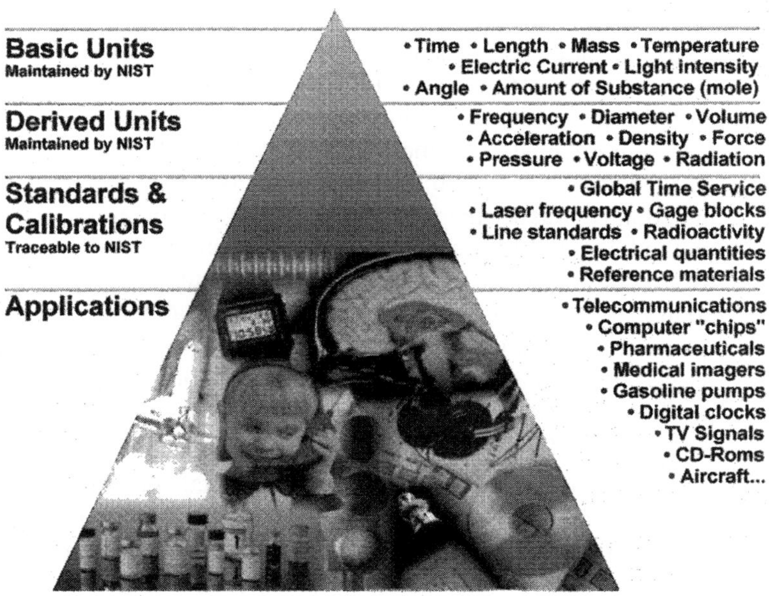

Figure 3.6. Areas for nanotechnology standards (courtesy M. Casassa, NIST).

- Other five-year goals include the following:
  - Batch-fabricated integrated measurement and lithography systems
  - Further investigation into top-down/bottom-up fabrication
  - Non-SPM probes that use electrons, ions, etc. (atomic-scale electron microscopy)
  - Intelligent analysis systems for medical, environmental, and defense applications
  - In-situ, nondestructive monitoring techniques for submonolayer control of superlattice growth

3.4 SCIENTIFIC AND TECHNOLOGICAL INFRASTRUCTURE

Building new infrastructure to support development of new tools and experimental methods must take into account the following considerations:

- The development of new instrumentation for nanostructure measurements, especially the scanning probes, has and will depend critically on synergestic work between university/government researchers (new ideas) and industrial developers (commercial realization). A government investment strategy must encourage and reward multidisciplinary collaborations among these communities.
- The small amount of material in, and the tiny size of, nanostructures frequently requires the use of special, expensive facilities: high-intensity synchrotron radiation sources, thermal neutrons, and high-energy electron beams (lithography and high-resolution electron microscopy). Adequate support for these facilities is important, both to create them and to provide affordable access to visiting researchers.
- While the scanning probes can be sufficiently inexpensive and routine for single-investigator acquisition and usage, state-of-the-art utilization of the probes can require highly specialized knowledge and apparatus. There should be a reasonable number of scanning probe analytical centers where that kind of knowledge and apparatus are available to visiting researchers. Those centers

should also be expected to continue advancement of scanning probe capabilities.

- The need for a database of information on proximal probe instrumentation, recipes for sample and probe preparation, standards and calibration procedures, and image analysis algorithms is becoming critical. An Internet-based information exchange could make this knowledge available to all potential users.
- Instrumentation development is not highly valued in the United States. To achieve the sophisticated instruments of tomorrow, it will be necessary to build the interest, knowledge base, and skill level of today's students. Toward this end, it will be helpful to create scholarships and fellowships to attract high-caliber high school students and post-docs interested in instrumentation and nanofabrication.

3.5 R&D INVESTMENT AND IMPLEMENTATION STRATEGIES

- Significantly increase investment for development of nano-instruments/tools that are low-cost, user-friendly, accurate, and reliable and user facilities that enable nanotechnology development. Early investment in instrumentation will yield benefits in all aspects of nanostructure science and technology.
- Foster industry-university-national laboratory cooperation in developing and commercializing nanoinstruments and tools.
- Ensure adequate support for high-performance beam sources (synchrotron light, neutron, and electron beams) for analytical facilities that provide affordable, state-of-the-art capabilities to the research community.

3.6 PRIORITIES AND CONCLUSIONS

The advancement of nanoscale science and technology can be facilitated by the development of nanoscale measurement instruments with improved capability. A major priority is to extend research instrument capability into low-cost, accurate, and reliable systems that can be used by researchers to explore new phenomena and to characterize fully nanostructured materials. Enabled by this capability, nanoscience advances will rapidly transition to applications in healthcare, food safety, environmental safety, law enforcement, and national security.

3.7 EXAMPLES OF CURRENT ACHIEVEMENTS AND PARADIGM SHIFTS

Several examples of new instruments and their utilization at nanoscale are presented below, including manipulation of single molecules and nanotubes, near-field optical and surface force microscopy, and observed nanoscale images on surfaces.

3.7.1 Single Molecule Manipulation and Measurement

Contact person: James Murday, Naval Research Laboratory

Tools to manipulate and measure single-molecule properties provide critical capabilities:

- Biology, medicine and healthcare will be revolutionized by the ability to manipulate the chemical/physical basis of living systems originating in the behavior of molecules at nanometer scales (see Chapter 10).
- As miniaturization continues, electronic structures will reach molecular sizes; carefully positioned single molecules can provide needed properties (see Chapter 8).
- Structural polymers, adsorbents, and supramolecular catalysts (e.g., enzymes) depend on molecular folding, shape, and reconfiguration (see Chapter 7) and can be designed for greater efficacy.

In the past, measurements of molecular behavior were necessarily ensemble averages; it was not possible to probe an individual molecule. While averaging techniques are very powerful, they mask detailed information necessary to fully understand the properties of matter, and more importantly, to enable full exploitation of the molecular behavior. The revolutionary advances in instrumentation featured in this chapter are providing exciting entrees into the single molecule world. Examples follow:

- *Carbon nanotubes.* Early theory predicted outstanding electrical and mechanical carbon nanotube properties that are now confirmed by measurements of individual nanotubes (Figure 3.7).

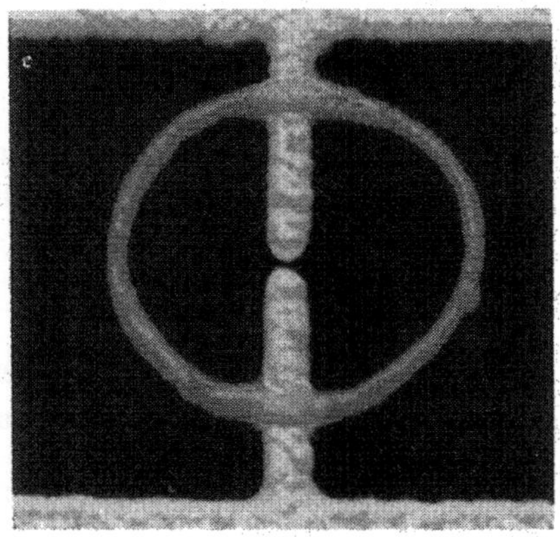

Figure 3.7. Carbon nanotubes circling electrical contacts (reprinted by permission from Dekker 1999, ©1999 American Institute of Physics).

- *Molecular recognition.* Much of biochemistry, including the immunoresponse critical to health, depends on molecules recognizing and binding to specific sites. Direct force-displacement measurements on bound molecules are now possible (Figure 3.8). This has led to a revolutionary approach to molecular detection—the force discrimination assay—where the recognition force between two biomolecules (antibody/antigen or complementary DNA strands) provides highly selective and sensitive detection.

- *Molecular motors.* Molecular motors are responsible for DNA transcription, cellular transport, and muscle contraction. The new microfabricated tools enable us to isolate, understand, and exploit these motors as new actuators for nanoelectromechanical tools—much smaller versions of microfabricated tools. This may lead to artificial biological devices, embedded in the body and powered by the same ATP that fuels normal body processes (Figure 3.9).

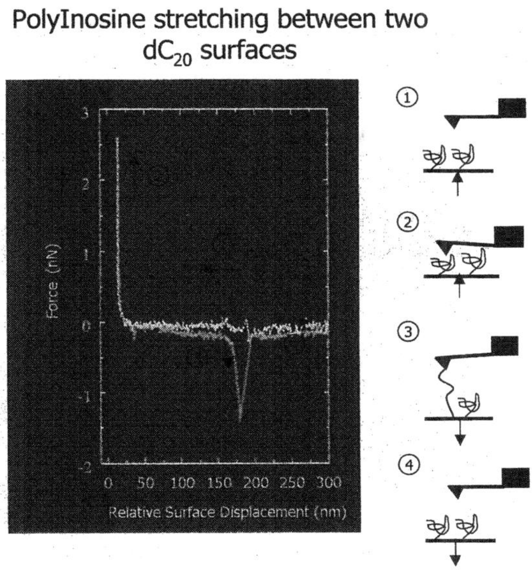

Figure 3.8. Force microscope measurement of complementary DNA binding (reprinted with permission from Baselt et al. 1996, ©1996 American Vacuum Society).

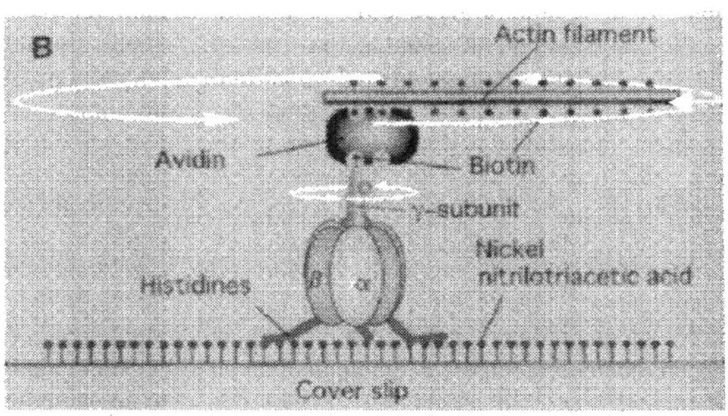

Figure 3.9. F_1-ATPase with actin filament mounted on a glass substrate (reprinted by permission from *Nature*, Noji et al. 1997, ©1997 Macmillan Magazines Ltd.).

- *Molecular folding*. A fundamental research problem in biochemistry is protein folding: how does a protein "know" its final configuration and achieve it quickly? Folding of structural polymers (e.g., crystallization, lamella formation) presents similar quandaries. A plethora of new techniques are providing direct molecular measurements of folding forces and dynamics, including optical tweezers and others (Figure 3.10).

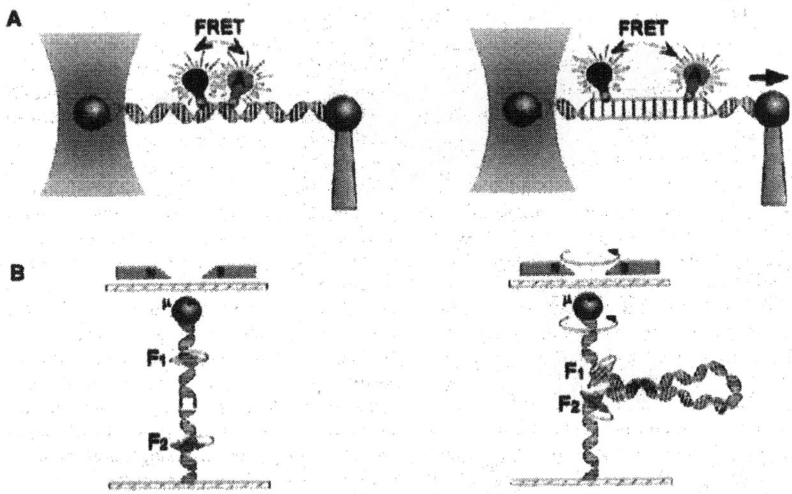

Figure 3.10. Optical tweezer and magnetic bead manipulation coupled with fluorescent probes (FRET) (reprinted with permission from Weiss 1999, ©1999 American Association for the Advancement of Science).

3.7.2 Nanomanipulator Inside a Scanning Electron Microscope

Contact Person: Rod Ruoff, Washington University, St. Louis

Innovations in manipulation and measurement of nanostructures are largely based in university and government laboratories; industry pays close attention to their discoveries and commercializes those that are most promising. As an example, a university-industry interaction between Washington University in St. Louis and Zyvex, a small business, has led to a new tool for manipulating nanoscale objects while simultaneously imaging with a SEM (illustrated in Figure 3.4 above). With this device, pulling, bending, and buckling of nanotubes into the third dimension are possible. The manipulator features a wide translation range, reasonable precision, small size, low-cost, and rapid assembly. Coarse 3-axis linear motions up to 6 mm and single-axis

360 degree rotational probe motion are provided by vacuum-prepared stainless steel stages driven by similarly prepped piezo actuators. An integral X-Y stage guides motion parallel to the plane of the SEM stage, and a separate Z-axis stage is used for motion along the SEM beam axis. Rotational motion normal to the beam is accomplished using a picomotor rotating actuator mounted atop the Z-stage. A four-quadrant piezo tube serves both as a support for the rotating tip and as a fine motion actuator in order to provide continuous motions augmenting the picomotor stepper action. Angular step sizes of <0.02 degrees with a maximum rotation rate of ~20 degrees/s, and spatial resolution of the piezotube of better than 0.1 nm are achieved.

Figure 3.11 shows SEM images of a single multiwalled carbon nanotube being stress-loaded after it has been attached across two atomic force microscope cantilevers; the imaging enables and confirms the attachment of a single tube. It also provides direct visual observation of tube dynamics.

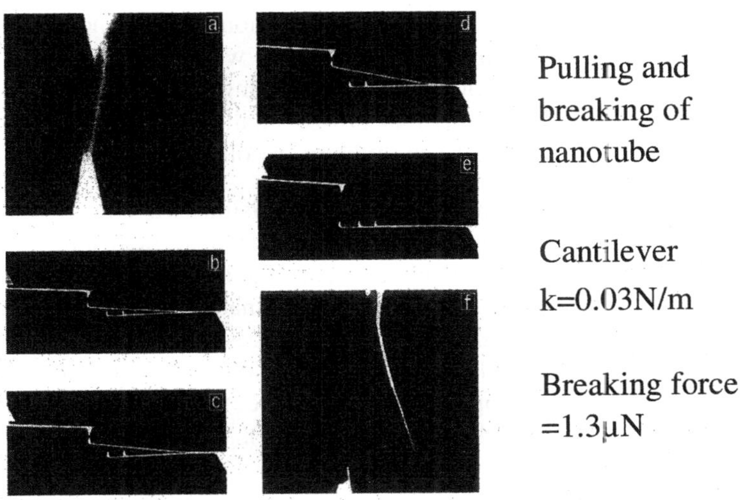

Figure 3.11. SEM images of a single multiwalled carbon nanotube being stress loaded and breaking away after it has been attached across two atomic force microscope cantilevers (Yu et al. 1999; reproduced by permission).

3.7.3 Multifunctional, Combined Near-Field and Surface Force Microscopes

Contact persons: Daniel van der Weide, University of Delaware, and James Murday, Naval Research Laboratory

Optical microscopy has been an essential tool in the scientific arsenal for centuries. Since the middle of the 1800s, the diffraction limit has constrained the resolution of optical images to the wavelength of light—about 0.5 micron in the visible spectrum. Development of scanning tunneling microscopy and atomic force microscopy in the 1980s provided imaging with three orders of magnitude better resolution. However, the basic physics in every form of microscopy limits what it measures. STM is predicated on electron tunneling; its images are defined by tunneling physics or by relaxation processes associated with the injected low-energy electrons. AFM has a broader range of capabilities; it can respond to a wide range of forces between tip and substrate—magnetic, Coulombic, dispersive, friction, core repulsion, etc. Optical imaging would complement STM/AFM images. Diffraction is a far field radiation effect; near field microscopy avoids the diffraction limit by working close to the sample. Several variants of near field scanning optical microscopy (NSOM) have been developed that utilize small apertures and/or tip antennae. Demonstrated visible light image resolution is ~10 nm.

Near field microscopes are not limited to visible wavelengths. A recent innovation has been the combination of near field and force microscopes (Figure 3.12). A miniaturized coaxial cable is fabricated onto a force microscope cantilever, terminating at a tip with nanometer dimensions. This geometry produces tiny probes with no cutoff frequency, is shielded to limit Coulomb interactions, and simultaneously probes topography (via force) and time varying electric fields (up to several GHz via near field).

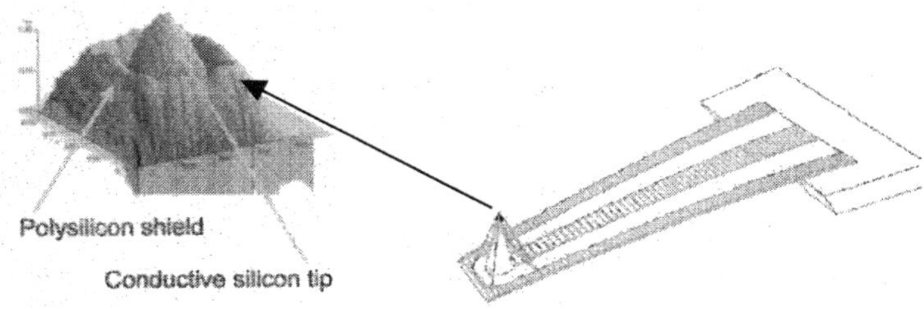

Figure 3.12. Near field antenna probe (after van der Weide and Neuzil 1996; reprinted by permission; ©1996 American Vacuum Society).

As an illustration, the image of a non-linear transmission line with ~100 nm topology is shown in Figure 3.13, along with the 30 ps waveforms detected at the specified point. This tiny near field antenna probe can operate in several modes: detection, excitation, reflection and transmission. It is a powerful new approach to the study of items as diverse as millimeter-wave electronic circuits and nerve cells.

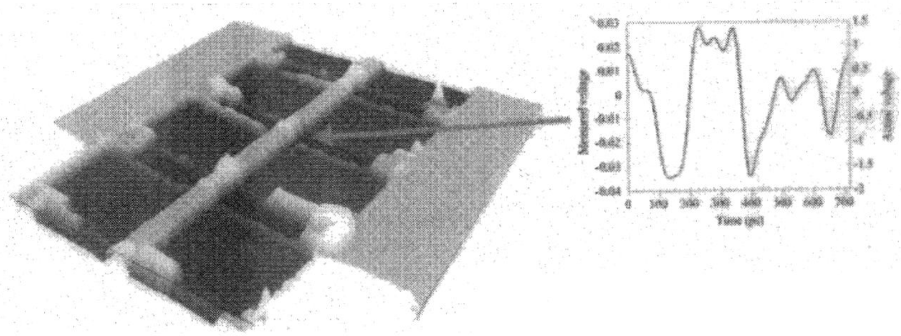

Figure 3.13. Near field antenna correlated measurement of topography and waveform (reprinted by permission from van der Weide 1997; ©1997 American Institute of Physics).

3.7.4 Image of Nanostructures on Surfaces

Contact person: P. West, ThermoMicroscopes

Figure 3.14 shows a monolayer of red blood cells on a mica substrate.

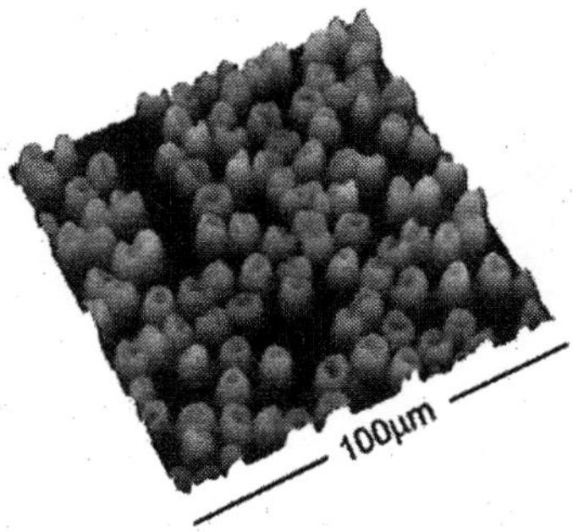

Figure 3.14. STM image and measurement of red blood cells (courtesy of L. McDonnell).

3.8 REFERENCES

Ade, H., W. Yang, S.L. English, I.J. Hartman, R.F. Davis, R.J. Nemanich, V.N. Litvinenko, I.V. Pinayev, Y. Wu, and J.M.J. Madey. 1999. A free electron laser-photoemission electron microscope system-(FEL-PEEM). *Surf. Rev. Lett.* 5 (6): 1257-68.

Bard, A.J., F.R.F. Fan, D.T. Pierce, P.R. Unwin, D.O. Wipf, and F.M. Zhou FM. 1991. Chemical imaging of surfaces with the scanning electrochemical microscope. *Science* 254:68-74.

Baselt, D.R., G.U. Lee, and R.J. Colton. 1996. A biosensor based on force microscope technology. *J. Vac. Sci. Technol. B* 14:789-793.

Baselt, D.R., G.U. Lee, M. Natesan, S.W. Metzger, P.E. Sheehan, and R.J. Colton. 1998. A biosensor based on magnetoresistance technology. *Biosensors and Bioelectronics* 13:731-739.

Bauer, E. 1990. Low energy electron microscopy. In *Chemistry and physics of solid surfaces VIII,* ed. R. Vanselow and R. Howe, 243-312. Berlin: Springer.

Bell, L.D. and W. J. Kaiser. 1996. Ballistic electron emission microscopy: A nanometer-scale probe of interfaces and carrier transport. *Annu. Rev. Sci.* 26:189-222.

Betzig E., J.K. Trautman, T.D. Harris, J.S. Weiner, and R.S. Kostelak. 1991. Beating the diffraction barrier: Optical microscopy on a nanometer scale. *Science* 251:1468-1470.

Bian, R.X., R.C. Dunn, and X.S Xie. 1995. Single molecule emission characteristics in near field microscopy. *Phys. Rev. Lett.* 75:4772-4775.

Binnig G, C.F. Quate, and C. Gerber. 1986. Atomic force microscope. *Phys. Rev. Letts.* 56:930-933.

Binnig G., H. Rohrer, C. Gerber, and E. Weibel. 1982. Surface studies by scanning tunneling microscopy. *Phys. Rev. Letts*. 49:57-61.

Birringer, R. 1994. Structure of nanostructured materials. In *Nanophase Materials,* ed. G.C. Hadjipanayis and R.W. Siegel. Amsterdam, Netherlands: Kluwer Academic Publishers.

Bürgi, L., O. Jeandupeux, A. Hirstein, H. Brune and K. Kern. 1998. Confinement of surface state electrons in Fabry-Pérot resonators. *Phys. Rev. Lett.* 81:5370-5373.

Chen, D.R. and D.Y.H. Pui. 1999. A high-efficiency, high-throughput unipolar aerosol charger for nanoparticles. *Journal of Nanoparticle Research.*

Chen, D.R., D.Y.H. Pui, D. Hummes, H. Fissan, F.R. Quant, and G.J. Sem. 1998. Design and evaluation of a nanometer aerosol differential mobility analyzer (nano-DMA). *Journal of Aerosol Science* 29:497-509.

Cohen, H., T. Maniv, R. Tenne, Y. Rosenfeld Hacohen, O. Stephan, and C. Colliex. 1998. Near-field electron energy loss spectroscopy in nanoparticles. *PRL* 80:782-785.

Cohen, M., J.M. Poate, and J. Silcox. 1993. *Atomic imaging and manipulation (AIM) for advanced materials*. Report of the NSF Panel on Atomic Resolution Microscopy, April.

Colton, R.J., G. U. Lee, and L.A. Chrisey. 1994. Direct measurements of the interaction of single strands of DNA with the atomic force microscope. *Science* 266:771-773.

Cowley, J.M. and J. Liu. 1993. Contrast and resolution in REM, SEM and SAM. *Surface Science* 298:456-467.

Dekker, C. 1999. Carbon nanotubes as molecular quantum wires. *Physics Today,* May, 22-28.

Despont, M., J. Brugger, U. Dreschsler, U. Durig, W. Haberle, M. Lutwyche, H. Rothuizen, R. Stutz, R. Widmer, H. Rohrer, G. Binnig, and P. Vettiger. 1999. *IEEE International Micro Electro Mechanical Systems Technical Digest 1999*:564. 17-21 January, Orlando, FL. Piscataway, NJ: IEEE.

Edelstein, A.S. and R.C. Cammarata, eds. 1996. Characterization of nanostructured materials. Part 5 in *Nanomaterials: synthesis, properties and applications.* London: Institute of Physics Publishing.

Edgerton, R.F. 1996. *Electronic energy-loss spectroscopy in the electron microscope.* 2d ed. Plenum Press.

Eigler, D.M., C.P. Lutz, and W.E. Rudge. 1991. An atomic switch realized with the scanning tunneling microscope. *Nature* 352(6336) (Aug 15): 600-603.

Gaub, H.E., M. Rief, M. Gautel, F. Oesterhelt, and J.M. Fernandez. 1997. Reversible unfolding of individual titin immunoglobulin domains by AFM. *Science* 276:1109-1112.

Gimzewski, J.K. and C. Joachim. 1999. Nanoscale science on single molecules using local probes. *Science* 283:1683-1688.

Ho, W., B.C. Stipe, and M.A. Rezaei. 1999. Localization of inelastic tunneling and the determination of atomic-scale structure with chemical specificity. *Phys. Rev. Lett.* 82:1724-1727.

Hobbs, P.C.D., D.W. Abraham, and H.K. Wickramasinghe. 1989. Magnetic force microscopy with 25 nm resolution. *Appl. Phys. Letts.* 55:2357-2359.

Hubert, A. and R. Schaefer. 1998. *Magnetic domains*. Berlin: Springer.

Kalbitzer, S., C. Wilbertz, and T. Miller. 1993. Intense focussed ion beams for nanostructurisation. In *Nanolithography: A borderland between STM, EB, IB and X-ray lithographies*. Vol. 264 of *NATO ASI series E: Applied sciences*, ed. M. Gentili, C. Giovannella, and S. Selci. Dordrecht: Kluwer Academic Publishers.

Kellermayer, M.S.Z., B. Smith, H.L. Granzier, and C. Bustamante. 1997. Folding-unfolding transitions in single titin molecules characterized with laser tweezers. *Science* 276:1112.

Kizuka, T., N. Tanaka, S. Deguchi, and M. Naruse. 1998. Time-resolved high-resolution transmission electron microscope using a piezo-driving specimen holder for atomic-scale mechanical interaction. *Micro. Microanal.* 4:218.

Knoll, B. and F. Keilmann. 1999. Near-field probing of vibrational absorption for chemical microscopy. *Nature* 399 (13 May): 134-137.

Kopp, M.U., A.J. deMello, and A. Manz. Chemical amplification: Continuous-flow PCR on a chip. *Science* 280:1046-1048.

Kurian, K.M., C.J. Watson, and A.H. Wyllie. 1999. DNA chip technology. *J. Pathology* 187:267-271.

Lang, H.P., R. Berger, F. Battiston, J.-P. Ramseyer, E. Meyer, C. Andreoli, J. Brugger, P. Vettiger, M. Despont, T. Mezzacasa, L. Scandella, H.-J. Guentherodt, Ch. Gerber, and J.K. Gimzewski. 1998. A chemical sensor based on a micromechanical cantilever array for the identification of gases and vapors. *Applied Physics A*. 66(7):S61-S64.

Leinhos, T., M. Stopka, and E. Oesterschulze. 1998. Micromachined fabrication of Si cantilevers with Schottky diodes integrated in the tip. *Appl. Phys. A.* 66:S65-S69.

Lemieux, B. A. Aharoni, and M. Schena. 1998. Overview of DNA chip technology. *Molecular Breeding* 4:277-289.

Luo, K., Z. Shi, J. Varesi, and A. Majumdar. 1997a. Sensor nanofabrication, performance and conduction mechanisms in scanning thermal microscopy. *J. Vac. Sci. Technol. B* 15(2):349-360.

Luo, K., R.W. Herrick, A. Majumdar, and P. Petroff. 1997b. Scanning thermal microscopy of a vertical-cavity surface-emitting laser. *Appl. Phys. Lett.* 71(12):1604-1606.

Majumdar, A, J.P. Carrejo, and J Lai. 1993. Thermal imaging using the atomic force microscope. *Appl. Phys. Lett.* 62:2501-2503.

Mankos, M., M.R. Scheinfein, and J.M. Cowley. 1996. Quantitative micromagnetics at high spatial resolution using far-out-of-focus STEM electron holography. *Advances in Imaging and Electron Physics* 98:323-426.

Mehta, A.D., M. Rief, J.A. Spudnich, D.A. Smith, and R.M. Simmons. 1999. Single-molecule biomechanics with optical methods. *Science* 283:1689-1695.

Miller, S.A., K.L. Turner, and N.C. MacDonald. 1997. Microelectromechanical scanning probe instruments for array architectures. *Rev. Sci. Instrum.* 68:4155-4162.

Minne S.C., G. Yaralioglu, S.R. Manalis, J.D. Adams, J. Zesch, A. Atalar and C.F. Quate. 1998. Automated parallel high-speed atomic force microscopy. *Appl. Phys. Lett.* 72:2340-2342.

Noji, H., R. Yasuda, M. Yoshida and K. Kinosita Jr. 1997. Direct observation of the rotation of F1-ATPase. *Nature* 386:299-302.

Noy, A., D.V. Vezenov, and C.M. Lieber. 1997. Chemical force microscopy. *Ann. Rev. Materials Science* 27:381-421.

Ohnishi, H., Y. Kondo, and K. Takayanagi. 1998. Quantized conductance through individual rows of suspended gold atoms. *Nature* 395:780.

Poncharal, P., Z.L. Wang, D. Ugarte, and W.A. de Heer. 1999. Electrostatic deflections and electromechanical resonances of carbon nanotubes. *Science* 283:1513.

Sato S. and H. Inaba. 1996. Optical trapping and manipulation of microscopic particles and biological cells by laser beams. *Opt. Quant. Electron.* 28 (1): 1-16.

Stöhr, J., Y. Wu, B.D. Hermsmeier, M.G. Samant, G.R. Harp, S. Koranda, D. Dunham, and B.P. Tonner. 1993. Element-specific magnetic microscopy with circularly polarized X-rays. *Science* 259:658.

Taylor II, R.M., W. Robinett, V.L. Chi, F.P. Brooks, Jr., W.V. Wright, R.S. Williams, and E.J. Snyder. 1993. The nanomanipulator: A virtual-reality interface for a scanning tunneling microscope. *Computer Graphics* (Proceedings of SIGGRAPH 93), August, 127-134.

Tomkiewicz, M. 1996. Surface characterization of nanostructured systems. In *Semiconductor nanoclusters: Studies in surface science and catalysis,* Vol 103, eds. R.V. Kamat and D. Miesel. Elsevier Science.

van der Weide, D.W. and P. Neuzil. 1996. The nanoscilloscope: Combined topography and ac field probing with a micromachined tip. *J. Vac. Sci. Technol. B* 14(6):4144-4147.

van der Weide, D.W. 1997. Localized picosecond resolution with a near-field microwave/scanning-force microscope. *Appl. Phys. Letts.* 70(6):667.

Wada, Y. 1997. Atom electronics: A proposal of atom/molecule switching devices. *Surface Science* 386:265-278.

Weisendanger, R. 1994. *Scanning probe microscopy and spectroscopy: Methods and applications.* Cambridge University Press.

Weiss, S. 1999. Fluorescence spectroscopy of single biomolecules. *Science* 283:1676-1683.

Wickramasinghe, H.K. 1989. Scanned-probe microscopes. *Scientific American.* 261 (Oct.): 98-105.

Williams, C.C., J. Slinkman, W.P. Hough, and H.K. Wickramasinghe. 1989. Lateral dopant profiling with 200 nm resolution by scanning capacitance microscopy. *Appl. Phys. Letts.* 55:1663-1664.

Yamamoto, N. 1997. Development of micro-fluorescent and diffracted X-ray spectrometer with a fine focused X-ray beam and its application of ULSI microanalysis. *Advances in X-Ray Analysis* 39:165-170.

Yazdani, A., B.A. Jones, C.P. Lutz, M.F. Crommie, D.M. Eigler. 1997. Probing the local effects of magnetic impurities on superconductivity. *Science* 275:1767-1770.

Yu, M.F., M.J. Dyer, G.D. Skidmore, H.W. Rohrs, X.K. Lu, K.D. Ausman, J.R. Von Ehr, and R.S. Ruoff. 1999. 3-dimensional manipulation of carbon nanotubes under a scanning electron microscope. *Nanotechnology.* 10(3):244-252. Bristol, U.K.: IOP Publishing Ltd.

Chapter 4

SYNTHESIS, ASSEMBLY, AND PROCESSING OF NANOSTRUCTURES

Contact persons: M. Tirrell, University of California, Santa Barbara; A. Requicha, University of Southern California; S. Friedlander, University of California, Los Angeles; G. Hagnauer, Army Research Laboratory

4.1 VISION

Synthesis and processing of nanostructures will employ a diverse array of material types—organic, inorganic, and biological—well beyond examples already realized. The driving forces will be creativity, applications, opportunities, and economics in broad areas of science, medicine, and technology. Increasing emphasis will be placed on synthesis and assembly at a very high degree of precision, achieved through innovative processing. The result will be control of the size, shape, structure, morphology, and connectivity of molecules, supermolecules, nano-objects and nanostructured materials and devices. Integration of top-down physical assembly concepts with bottom-up chemical and biological assembly concepts may be required to create fully functional nanostructures that are operational at mesoscopic scales. The combination of new nanoscale building blocks and new paradigms in assembly strategies will provide nanostructured materials and devices with new, unprecedented capabilities limited only by our imagination.

4.2 CURRENT SCIENTIFIC AND TECHNOLOGICAL ADVANCEMENTS

Recent Scientific Advances

Synthesis of Individual Building Blocks

Polymeric materials, dendrimers, and block copolymers. The last decade has seen tremendous advances in the preparation of organic building blocks of considerable complexity (Matthews et al. 1998; Stupp et al. 1997; Tomalia 1994). The discovery of a new topology for polymers, dendrimers, has led to an exciting new class of nanoscale component, with interesting optical and mechanical properties. Precise nanoscale architectures ranging between 10 and 100 nm have been successfully synthesized. These constructions involve the reaction of an excess of dendrimer shell reagent with a reactive dendrimer core reagent. The new compositions are referred to as tecto (dendrimer) core-shell molecules. These molecules have demonstrated potential as unique nanoscale reactors, intermediates for new coatings/controlled delivery, compatibilizers, and building blocks for higher order nanoscale constructions. There have also been steady advances in engineering new phases using block copolymers; the recent development of tri-component block copolymer is noteworthy in this regard.

Nanocrystals. There has been significant progress made in the preparation of nanocrystals in recent years (Brus 1996; Martin 1996). Many common materials, such as metals, semiconductors, and magnets, can be prepared as nanocrystals, using colloidal chemistry techniques. The concepts of ligand exchange and surface derivatization have been well developed, and these methods permit nanocrystals with narrow size distribution (typically 5-15% variation in diameter) to be isolated and then used further as chemical reagents. This field has been aided greatly by improved understanding of size-dependent scaling laws, which have emerged from fundamental studies in chemical physics and condensed matter physics. The fact that a simple property like light emission depends so strongly upon size in semiconductors has greatly facilitated the development of reliable preparations. The same size dependence has also led to a wide range of applications in unexpected areas, such as in biological tagging (Chan and Nie 1998; Bruchez et al. 1998).

Nanotubes and rods. The exciting discovery of the fullerenes was followed closely by the discovery of nanotubes of carbon (Terrones et al. 1999). Nanotubes show tremendous promise as building blocks for new materials. Because of their topology, nanotubes have no dangling bonds, and so despite being very small, they do not exhibit "surface effects." As a consequence, individual nanotubes exhibit nearly ideal electrical, optical, and mechanical properties. Nanorods are also under extensive development and investigation.

Nanoparticle structures. Controlled particle formation is an important synthetic route to nanoscale building blocks relevant to many technologies from ceramics to pharmaceuticals. Some interesting new nanoparticle structures are composed of chain-like arrays of nanoparticles of relatively low coordination number. There are two main types: agglomerates (or aggregates) and aerogels. In particular, these structures can be characterized by their morphology (for example, fractal dimension and coordination number) and the energies of the bonds that hold the primary (individual) particles together.

Processing of Nanostructures

Assembly. The development of self-assembly methodology, which is the archetypal bio-inspired synthesis route, has greatly expanded the methods of construction of nanostructures. In the design of complex materials such as electrical devices, we currently rely on our ability to create designed patterns lithographically. New ways of bonding, assembly, and linking macromolecules and nano-objects have been developed that are based on interactions that are both more complex and individually weaker (e.g., steric, electrostatic, hydrophobic, and hydrogen bonding) than the classical electronic bond. Multiple bonding interactions are often needed to stabilize complex nanostructures. These interactions are the basis for coding information into nanostructures. In the last decade, nanoscale objects such as nanoparticles or nanocrystals have been assembled into periodic arrays, or supercrystals. Such arrays exhibit novel optical and electrical characteristics. Several proposals have been put forward for how to pattern nanocrystals and nanotubes using biological molecules (Mucic et al. 1998; Alivisatos et al. 1996; Braun et al. 1998).

Templated growth of mesoporous materials. In the last decade, tremendous advances have occurred in the preparation of mesoporous inorganic solids (Antonelli and Ying 1996). The initial work showed that it is possible to use organic surfactant molecules to prepare a complex pattern. That pattern can serve as the template for the formation of an inorganic phase. This has led to many exciting discoveries in chemical synthesis and to immediate practical advances in catalysis. Nanoporous media science (the control of void space) has advanced in some very important ways. For example, new scaffolds and matrices for tissue repair and engineering have been realized, and a large range of tailored porous catalysts and membranes, such as Mobil's MCM-41, have achieved commercial success. In another example, Nylon-6 nanocomposite with only two volume percent clay nanoparticles has a heat deformation temperature of 150°C, as opposed to 60°C for traditional Nylon-6.

Direct structuring. The ability to direct the assembly and organization of materials with nanomanipulation and nanolithography, based, for example, on scanning microprobe techniques, has achieved directed assembly and structuring of materials at the molecular level. New methodologies in this area include 3-D printing and various forms of soft lithography.

Nanoimprint lithography. Nanoimprinting will allow for patterning at scales up to 10 nm on large surfaces and with a relatively low cost (see section 4.7.2).

Recent Technological Advances

One key to advanced technology emerging over the past decade has been nanomaterials. The aggressive advance of smart materials, solid state devices, and biomimetic technologies and the concurrent push towards miniaturization are making the understanding and development of materials on the nanometer level critical and are encouraging the design of nanoscale structure and functionality into materials systems. The focus on nanostructuring of materials systems has been further sharpened by the need to develop materials having novel and/or enhanced properties without resorting to new synthetic chemistries with the associated environmental and cost issues. Enhancements in mechanical performance, wear resistance, integrity under thermal stress, flammability, and transport properties have *all* been linked to nanostructure in materials systems within the past five years, demonstrating that the technology has reached a level of maturity where it is ripe for exploitation in systems demanding both high performance and reliability.

A very important technological advance in recent years has been in the area of large-scale, reliable production of uniform, nano-sized particles. This has been particularly important in the high-performance ceramic materials and the pharmaceutical areas, where materials properties, through defect control, drug delivery, and control of uptake, have been favorably influenced by nanoparticle production. Aerogels are normally fabricated by condensed phase (sol-gel) methods, even though the final product is a gas/solid system. Recently, aerogel-like structures have been fabricated directly by gas-phase processes without passing through the sol-gel state. This could lead to less expensive fabrication processes, use of a wider range of materials in aerogel fabrication, and excellent control of multilayer deposition processes, with applications in magnetic (giant magnetoresistance, GMR) and optical devices. Soft lithography and nanoimprinting have been developed and identified as low-cost patterning approaches, with several new applications on the horizon. Nanostructured zeolite catalysts can be tailored to perform oxidation reactions more efficiently than enzymes. While not strictly speaking a nanotechnology, the tremendous advances in organic electronics, such as organic light-emitting diodes, must be noted, since that field is highly likely to benefit from advances in organic nanoscale synthesis in the future.

Potential Impact

New nanostructured materials have the potential to significantly reduce production costs and the time of parts assembly, for example, in the automotive, consumer appliance, tooling, and container industries. The potential of significant reductions in weight due to these new materials as they are applied in the transportation industries will have great impact on energy consumption and the environment. Understanding nanoparticle formation is paying dividends in dealing with environmental issues such as atmospheric particulate formation as well.

Many fundamental phenomena in energy science, such as electron transfer and exciton diffusion, occur on the nanometer length scale. Thus, the ability to arrange matter, i.e., to inexpensively pattern and to develop effective nanostructuring processes, will be a vital asset in designing next-generation electronic devices, photovoltaics, and batteries.

Size and cost reduction due to advances in the design and manufacture of healthcare-related diagnostic systems has the potential to empower individuals to diagnose and treat diseases in their own homes, decentralizing the healthcare system.

Sensors based on nanotechnology will revolutionize healthcare (e.g., via remote patient monitoring), climate control, detection of toxic substances (for environment, defense, and healthcare applications), and energy consumption in homes, consumer appliances, and power tools.

The ability to assemble and interconnect nanoparticles and molecules at nanometer dimensions will enable the development of new types of nanoelectronic circuitry and nanomechanical machinery.

4.3 GOALS FOR THE NEXT 5-10 YEARS: BARRIERS AND SOLUTIONS

New Discoveries and Applications Anticipated in the Next Decade

There is broad opportunity in the next decade for synthesis and processing in applications at the interface with biology. Specific areas include biological synthesis using coded self-assembly and guided assembly using biorecognition capabilities of DNA and proteins; nanoparticles for drug delivery, gene therapy, and immunotherapy; and a wide range of biological probes and sensors. Increasing success is anticipated with bio-inspired processes that interface assembled nanostructures with biological systems.

High-throughput screening methods, that is, methods that measure properties or activities rapidly in spatially addressable ways, will be necessary in order for combinatorial chemistry methods to realize their full potential in new drug and materials development. New nanotechnology is both an enabler and a result of the development of new high-throughput screening. For example, robotics is expected to be very important in achieving these goals.

Nanotechnology and synthesis will open new frontiers in the design of catalysts and catalyst technology for the petroleum, chemical, automotive, pharmaceutical, and food industries. The design of catalyst supports commensurate with biological structures will be an important bridge between conventional and enzymatic catalysis. In fact, oxidation catalysis can be performed today more efficiently in a zeolite, "ship-in-a-bottle" catalytic complex than with natural enzymes. This is but one example of an entire array of anticipated future developments.

New discoveries are expected and needed in studies of single objects with nanoscale dimensions ranging in size from single molecules, clusters, and particles to organelles and cells. Researchers will learn more about the opportunities for and limits on the synthesis of large, precisely structured objects and clusters. Controlling purity and scale-up of products emanating from such precision syntheses is a major barrier that must and will be tackled in the near future. Many of the important properties of nanostructures depend on obtaining precise building blocks; means of creating and analyzing purity and homogeneity in such products are vitally needed. Furthermore, if production of these materials cannot be done at a sufficiently large scale, this will eventually limit utility in some applications.

While current microfluidics approaches will be effective for manipulating single objects on the scale of one micron or more, new techniques must be developed for single-object manipulation at smaller scales. The ability must be developed to do nanomanipulation in three dimensions to guide nanoassembly in bulk as well as on surfaces. There will be increasing interactions between nanoscale scientists and system designers. An important element of this interaction will be prototyping methods, an intermediate level of implementation between lab-scale demonstration and mass production.

The new nanomaterials will impact not only the performance of the most advanced computational and electronic devices, but also objects of daily use familiar to every consumer, such as cars, appliances, films, containers, and cosmetics.

Paradigm Changes

A significant paradigm shift is expected, owing to our improved ability to address, manipulate, and activate individual molecules and objects. Increasingly, important developments will be made with hybrid or nanocomposite materials, that is, combining very different materials systems (organic, inorganic, and biological) in one integrated structure.

The integration of nanotechnology in medicine, supported by fundamental science bridging the gap between nanotechnology and biology, will be critical in bringing the impact of nanotechnology to the attention of the public. This integration can be expected to bring about revolutionary changes in healthcare as well as advances in biology itself. Further detailed ideas on this are given in Section 4.7.6.

New computer architectures will require new approaches to synthesis and assembly, reflecting the idea (expressed at the IWGN workshop by Horst Stormer) that "wiring" may be more important than "wires" or "switches" in assembling functional nanostructures for computing. New methods for connecting elements of nanostructures will be required. The strain-directed assembly of nanoparticle arrays in a solid (Figure 4.7, after Kiehl et al. 1996), where a top-down physical process (lithographically defined surface structure) is integrated with a bottom-up chemical assembly process (strain and compositionally controlled precipitation), represents an example of a significant trend of integrating fabrication techniques that will be crucial for the fabrication and interconnection of wires and switches. Such integration techniques will provide the means for introducing functionality in the substrate that is coupled to functionality on the surface. A new form of information technology is emerging characterized by ubiquitous interaction with information and the physical world. Full implementation of this will require enormous numbers of sensors and actuators in addition to very small computers.

The education and training of young scientists and engineers in environments conducive to exploring these new paradigms will be essential. This will require some creative approaches to interdisciplinary education.

4.4 SCIENTIFIC AND TECHNOLOGICAL INFRASTRUCTURE

In the area of synthesis, assembly, and processing of nanostructures, several kinds of infrastructure are very important. State-of-the-art characterization tools underpin all efforts to synthesize and manufacture high-precision, high-purity substances; advanced characterization methods must be accessible to those doing this type of work. These include, but are not limited to, neutron, X-ray, and light-scattering tools (some of which require advanced sources such as reactors, spallation sources, and synchrotrons); surface and interface analytical tools; particle characterization; microscopy of all types; and rheological methods. Equally important are synthesis tools themselves, such as nanofabrication facilities. New synthesis facilities, particularly those that might make new materials widely available to a broader range of investigators, could advance the field significantly. Large-scale scientific computation facilities are very important in the design and characterization of nanostructures.

4.5 R&D INVESTMENT AND IMPLEMENTATION STRATEGIES

The guiding principles of R&D investment should be (1) support work that crosses all traditional boundaries, and (2) maintain appropriate balance between centers, teams, and single-investigator grants, and between basic science work and device/applications work. Boundaries to be crossed include those between traditional academic disciplines, between universities and industry, and between countries. Nanoscience and technology requires a spectrum of diversity of talent and approaches that cannot be achieved without crossing boundaries.

4.6 PRIORITIES AND CONCLUSIONS

The large variety of avenues in this broad area of research makes it a richly diversified area for investment objectives. Synthetic chemistry now has the most diverse set of research targets. Priority in future research should be given to projects with clearly articulated interdisciplinary tools. Research on synthesis should endeavor to link itself with research on scale-up or on advanced processing, or with research in fundamental biology. No one field has a monopoly on the tools that will be created to solve these problems: the more interdisciplinarity that can be brought to bear the better. Nanotechnology is perhaps the one field with the most to exploit from bringing all disciplines closer together. Scaleup of processes, and related chemical engineering research, are particularly neglected areas that are necessary to realize the full potential for effective synthesis, assembly, and processing of nanostructures.

4.7 EXAMPLES OF CURRENT ACHIEVEMENTS AND PARADIGM SHIFTS

4.7.1 Fullerenes and Nanotubes

Contact person: D.T. Colbert, Rice University

Fullerene nanotubes hold tremendous promise for numerous applications, owing to their remarkable materials properties, including strength, stiffness, toughness, chemical robustness, thermal conductivity, and perhaps most interestingly, electrical conductivity. Depending on their precise molecular symmetry, some nanotubes are semiconducting, while others exhibit truly metallic conductivity. This behavior, coupled with their nanoscale geometry, makes them ideal—perhaps unique—candidates for wires, interconnects, and even devices for true molecular electronics.

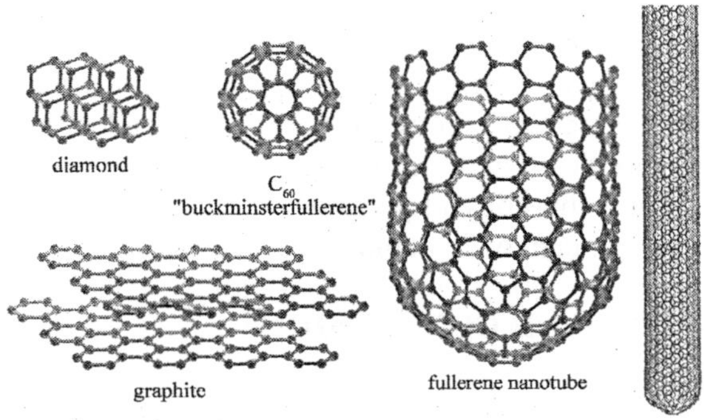

Figure 4.1. Fullerene nanotubes.

One application of nanotubes—as probe tips in scanning probe microscopy (Dai et al. 1996; Wong et al. 1998)—has already been developed. Many others, such as field-emission displays (Rinzler et al. 1995; de Heer et al. 1995); high-strength composites, and various electronic applications, are being pursued vigorously now, largely enabled by the discovery in 1995 (Guo et al. 1995; Thess et al. 1996) of the laser-vaporization process for producing single-wall nanotubes in high yield. In the three years since this breakthrough, a tremendous amount has been learned about the fundamental physical characteristics of fullerene nanotubes, mostly consistent with early expectations of extraordinary material properties. The gram quantities of nanotubes provided by the laser process, and now by the arc process as well, are enabling a period of research on chemical methods for manipulating and assembling short lengths of nanotubes (Liu et al. 1998; Chen et al. 1998). These are expected, in turn, to provide the enabling technologies for the applications exploiting the material properties discussed above. It must be stressed that the full realization of most applications exploiting these properties will be made possible by the very high degree of structural perfection exhibited by nanotubes. This *molecular* aspect of fullerene nanotubes permits us to develop chemical strategies for assembling them into useful structures, materials, and perhaps molecular electronic devices.

4.7.2 Nanoimprint Lithography

Contact person: S. Chou, Princeton University

Nanoimprint lithography (NIL) is a revolutionary approach to low-cost and high- throughput nanolithography (Chou 1998; Chou et al. 1996). NIL patterns a resist by physically deforming the resist shape with a mold (i.e., embossing), rather than by modifying the resist chemical structures with radiation as in a conventional lithography (Figure 4.2). This fundamental difference in principles frees NIL from many problems suffered in conventional lithography, such as diffraction limit, scattering, and chemistry. As a result (see also Figure 4.3), NIL can achieve sub-10 nm structures over large areas with low cost and high throughput—a feat currently unachievable using existing lithographies.

Successful development of NIL will bring a revolution to nanostructure research, because NIL will remove the key obstacle—

cost—to nanostructure commercialization and will make nanostructures easily accessible to everyone. To a great extent, one can compare the impact of NIL with that of personal computers, which have made computation so widely accessible. Therefore, NIL will not only impact future integrated circuit development, but will also impact many other disciplines, such as biology, chemistry, medicine, and materials, to name a few.

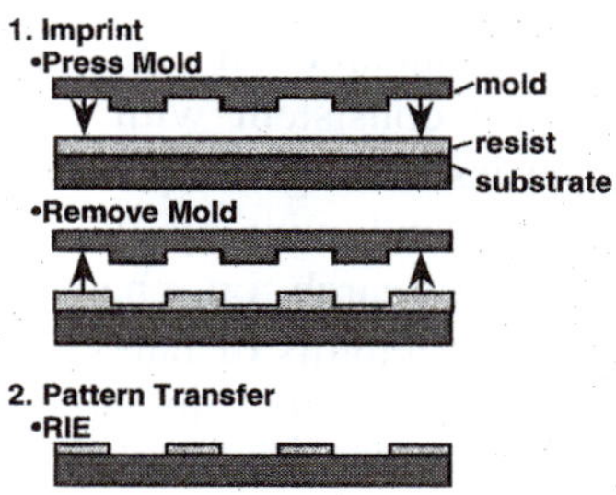

Figure 4.2. Schematic of nanoimprint lithography process: (1) imprinting using a mold to create a thickness contrast in a resist, and (2) pattern transfer using anisotropic etching to remove residue resist in the compressed areas (reprinted with permission from Chou et al. 1996, ©1996 American Association for the Advancement of Science).

Figure 4.3. SEM micrograph of a top view of 10 nm minimum diameter and 40 nm period holes imprinted into PMMA (60 nm deep) (reprinted with permission from Chou et al. 1997, ©1997 American Vacuum Society).

4.7.3 Lithographically Induced Self-Assembly

Contact person: S. Chou, Princeton University

Lithographically-induced self-assembly (LISA) is a recent discovery that will have a great impact on science and technology (Chou and Zhuang 1997, 1999). In LISA, a mask is used to induce and control the self-formation of periodic supramolecular pillar arrays in a thin polymer melt that was initially flat on a substrate. The mask was initially placed above the polymer film with a gap. The pillars, formed by rising against the gravitational force and surface tension, bridge the two plates. The boundary of the pillar array is precisely aligned to the bounding contour of the patterns on the mask (Figure 4.4). The principle for LISA is, although still unclear, fundamentally different from self-assembly by phase-separation and surface chemistry modification. It is believed that LISA is related to electrostatic forces and electrohydrodynamic instabilities (Chou and Zhuang 1999).

LISA opens up exciting new areas for fundamental scientific study and practical applications. Scientifically, understanding of the LISA principle requires combining several disciplines. Technologically, LISA offers a solution to the two long-sought goals: (a) precise control of the orientation and location of a self-assembled polymer structure, and (b) making the self-assembled features smaller than those of mask patterns (Figure 4.5). Furthermore, the LISA process should, in principle, be applicable to other polymers and perhaps even other single-phase materials, such as semiconductors, metals, and biological materials. The periodic arrays formed by LISA have many applications, such as memory devices, photonic materials, and new biological materials, to name a few. Finally, LISA offers a unique way to pattern polymer electronic and optoelectronic devices directly without using the detrimental photolithography process.

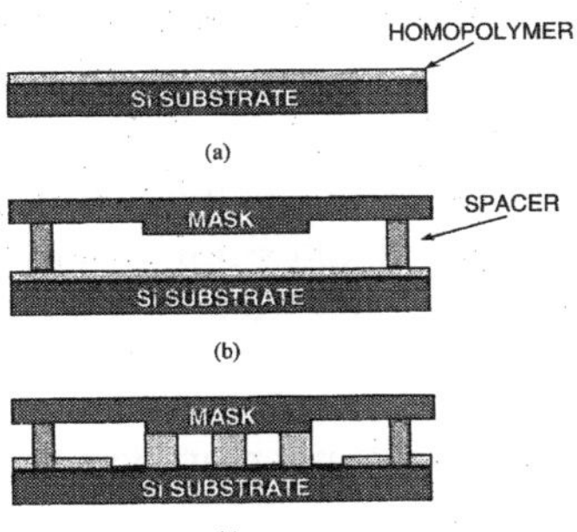

Figure 4.4. Schematic of lithographically-induced self-assembly (LISA). A mask is used to induce and control the self-formation of supramolecular pillar array in a thin polymer melt (reprinted with permission from Chou and Zuang 1999, ©1999 American Vacuum Society).

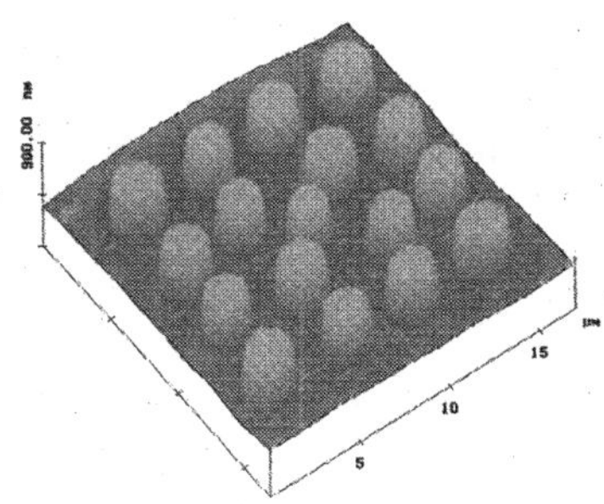

Figure 4.5. AFM image of PMMA LISA pillar array formed under a square pattern (reprinted with permission from Chou and Zuang 1999, ©1999 American Vacuum Society).

4.7.4 DNA-Directed Assembling: Potential for Nanofabrication

Contact Person: M.J. Heller, Nanogen

DNA chips and microarrays represent a technology which has immediate applications in genetic research and diagnostics. DNA array technology may also play a future role in enabling nanofabrication. DNA chips or arrays are devices in which different DNA sequences are arrayed in a microscopic format on a solid support (glass, silicon, plastic, etc.). DNA arrays can have anywhere from 100 to 100,000 different DNA sites (pixels) on the chip surface. Depending on the chip, the sites can range in size from 10 microns to over 100 microns (smaller sites are possible). Each DNA site can contain from 10^6 to 10^9 DNA sequences. In a DNA hybridization assay, the DNA array is contacted with a sample solution that contains the unknown target DNA sequences. If any of the sequences are complementary to those on the array, hybridization occurs and the unknown sequence is identified by its position on the array. A number of companies are now involved in the development of DNA chips and arrays, including Affymetrix, PE Applied Biosystems, HySeq, Nanogen, Incyte, Molecular Dynamics, and Genometrix. Present DNA chip devices will have applications in genomic research, pharmacogenetics, drug discovery, gene expression analysis, forensics, cancer detection, and infectious and genetic disease diagnostics.

Newer generations of electronically active DNA microarrays (under development by Nanogen) that produce controlled electric fields at each site may have potential applications for nanofabrication. These active microelectronic devices are able to transport charged molecules (DNA, RNA, proteins, enzymes), nanostructures, cells and micron-scale structures to and from any test site on the device surface. When DNA hybridization reactions are carried out, these devices are actually using electric fields to direct the self-assembly of DNA molecules at specified sites on the chip surface. These active devices are serving as semiconductor hosts or motherboards for the assembly of DNA molecules into more complex three-dimensional structures. The DNA molecules themselves have programmable and self-assembly properties and can be derivatized with a variety of molecular electronic or photonic moieties. DNA molecules can also be attached to larger nanostructures, including metallic and organic particles, nanotubes, microstructures, and silicon surfaces. In principle, active microelectronic arrays and DNA-modified components may allow scientists and engineers to direct self-assembly of two- and three-dimensional molecular electronic circuits and devices within the defined perimeters of larger silicon or semiconductor structures (Figure 4.6). Thus, electronically directed DNA self-assembly technology could encompass a broad area of potential applications from nearer term heterogeneous integration processes for photonic and microelectronic device fabrication to the longer term nanofabrication of true molecular electronic circuits and devices.

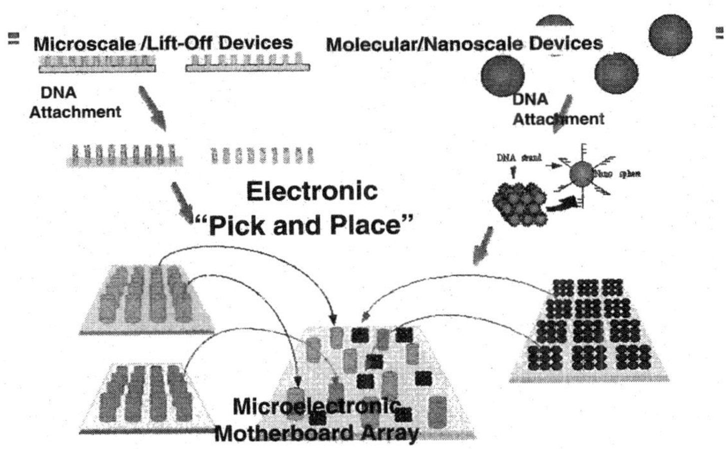

Figure 4.6. Directed nanofabrication on a chip (Nanogen, Inc.).

4.7.5 Strain-Directed Assembly

Contact person: R. Kiehl, University of Minnesota

The integration of top-down physical and bottom-up chemical or biological assembly methods will be crucial for the fabrication and interconnection of wires and switches. The strain-directed assembly of nanoparticle arrays in a solid (Figure 4.7), where a top-down physical process (lithographically defined surface structure) is integrated with a bottom-up chemical assembly process (strain and compositionally controlled precipitation), represents a step in this direction. More generally, the development of such techniques will provide the means for introducing functionality in the substrate that is coupled to functionality on the surface. Strain-directed assembly of arsenic precipitates in an AlGaAs/GaAs heterostructure is sketched in Figure 4.7. The horizontal positions of the 20 nm particles are controlled by the 200 nm surface stressors, while the vertical positions are confined to a 10 nm GaAs layer. One-dimensional arrays of closely spaced particles are formed along lines running into the plane (Kiehl et al. 1996).

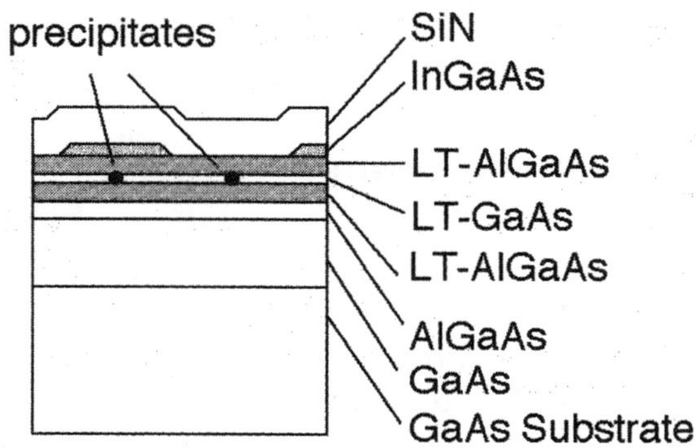

Figure 4.7. Strain-directed assembly.

4.7.6 Nanotechnology Synthesis and Processing in Drug and Gene Delivery

Contact person: K. Leong, Johns Hopkins University

Almost half of therapeutically useful drugs are hydrophobic. Administration of these water-insoluble drugs is problematic. The bioavailability of these drugs can be significantly enhanced by reducing the size of the drug particles to the nanoscale. Thus small enough to pass through the capillaries, the drug may even be administered via intravenous injection. The benefit to the pharmaceutical industry of this nanotechnology processing has been enormous.

Genetic medicine continues to hold exciting promise in the future of healthcare. A major challenge for successful gene therapy has been the development of safe and efficient gene vectors. While viruses in some cases can efficiently deliver exogenous genes to cells in vivo, the long-term safety of this approach remains a major concern. Non-viral vectors have been increasingly proposed as alternatives.

Nanoparticles composed of complexes between polycationic lipids or polycationic polymers with DNA have shown efficacy in many animal models. The lipid-DNA complexes are being tested in several clinical trials, notably the delivery of the CFTR gene to the lung airways for correcting the chloride transport defect that leads to cystic fibrosis. These DNA nanoparticles may potentially be the most practical vehicles for fulfilling the promise of genetic medicine.

Nanotechnology Synthesis and Processing in Drug/Gene Delivery

Producing drug particles down to the nanometer scale, uniform in size and distribution, non-aggregated in solution, and manufacturable in industrial scale remains a significant challenge. Continuing advances in nanotechnology, particularly the fundamental aspects, will be needed to meet this challenge. New nanosynthetic approaches may be needed to improve current techniques such as controlled crystallization, and improvements over the milling and scale-up processes will be important.

Nanotechnology may also help reach the hitherto elusive goal of active drug targeting. The "magic bullet" concept has mostly been tested on soluble complexes or targeting ligands conjugated to ill-defined particles. Limited success has been documented in the literature on delivering polymer-coated nanoparticles across the blood-brain barrier or increasing the lymphatic drainage of nanoparticles to target the lymph node. Advances in nanotechnology that can further reduce the size and reproducibly attach targeting ligands to the drug-loaded nanoparticles may improve the targeting efficiency. These nanoparticles may also be valuable tools for molecular biologists to study the cellular processes of receptor-mediated endocytosis and intracellular trafficking. A potentially important application of these nanoparticles may be altering the way an immunogen can be presented to the immune system of the host. An antigen adsorbed to or encapsulated in nanoparticles may be used to optimize the immune response in vaccine applications.

Current non-viral gene vectors are far from perfect. Ideally, DNA nanoparticles with controlled composition, size, polydispersity, shape, morphology, stability, encapsulation capability, and targetability would be needed to optimize the transfection efficiency in vivo. Scaling up of the DNA particle synthesis is also a serious challenge.

Only with significant advances in nanotechnology will the potential of these DNA nanoparticles be realized.

There is a strong need to expand the effort to investigate the fundamental aspects of nanosynthesis and processing related to drug and gene delivery. Nanosynthesis by complex coacervation remains an inexact science. A theoretical framework that can describe and predict the phase separation behavior of such polyelectrolytes will greatly aid the choice of polycationic carriers and the synthesis. A clear picture of the self-assembly of the DNA-polycation complex, such as the studies conducted on lipid-DNA complex, will help correlate the physico-chemical properties with the transfection efficiency. A detailed biological transport analysis of these nanoparticles will help define the mechanism and identify the rate-limiting steps of the transfection process. A better understanding of colloidal behavior in biological fluids will also facilitate the rational design of these nanoparticular drug and gene delivery systems.

4.7.7 Nanostructured Polymers

Contact person: S.I. Stupp, Northwestern University

Nano-sized polymers shaped as rounded bricks, cones, mushrooms, and plates have been prepared in laboratories and found to self-assemble into tubular, spherical, layered, and lamellar constructs, respectively. These new types of polymers may eventually be useful in applications ranging from sophisticated sensors to de-icing agents (Stupp 1998).

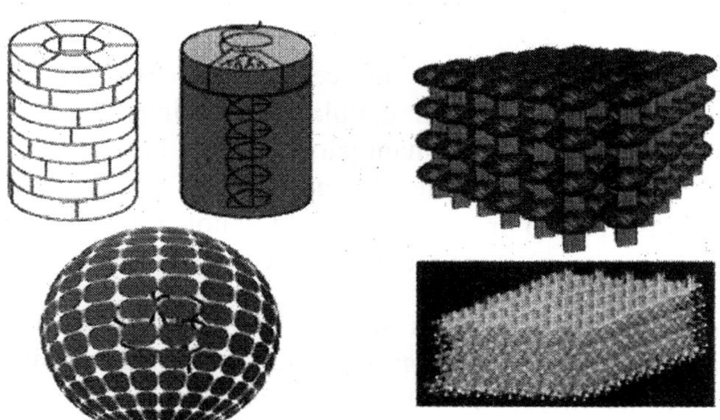

Figure 4.8. Nano-sized polymers self-assembling into functional structures.

4.7.8 Replication of Nanostructures by Polymer Molding

Contact persons: R.J. Celotta and G. Whitesides

A key element in the utilization of nanostructures for as many applications as possible is the ability to inexpensively mass-produce them. The technique of polymer molding, long used for replication of micron-sized structures in such devices as diffraction gratings, compact disks, and microtools, has now been shown to work on the nanoscale as well (Xia et al. 1997). Beginning with a master nanostructure, a mold is made using an elastomer such as polydimethylsiloxane (PDMS). The mold is then used to produce replicas in a UV-curable polymer such as polyurethane. As seen in Figure 4.9, which shows atomic force microscope images of the original master and a replica, high-quality reproduction is possible on a scale of tens of nanometers.

The demonstration that this replication process works on the nanoscale was carried out using a master pattern fabricated via a unique new process known as laser-focused atomic deposition (McClelland et al. 1993). In this process, a laser standing wave forms an array of atomic "microlenses." These concentrate chromium atoms as they deposit onto a surface, building nanoscale objects "from the bottom up" in a single step without the use of any resist.

Replication of the laser-focused chromium structure is only one example of the use of polymer molding on the nanoscale. For example, nanostructures with ~30 nm lateral dimensions have been produced based on gold patterns made using conventional lithography.

Given these demonstrations, it is now clear that a new tool is available for nanoscale fabrication, with a direct avenue to mass production. As applications for nanostructured materials continue to expand, this technology stands ready for implementation in a manufacturing setting, enabling the type of inexpensive production techniques that new technologies critically depend on.

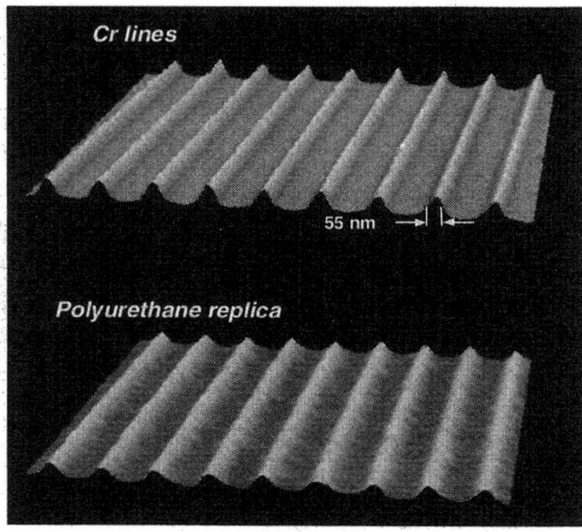

Figure 4.9. Soft lithographic nanostructure (courtesy J.J. McClelland, NIST).

4.7.9 Molecular Self-assembly

Contact person: M. Reed, Yale University

Figure 4.10 depicts a single molecule bridging the gap between two metallic contacts, forming the smallest and ultimate limit of an electronic device. The illustration points to a potentially powerful new fabrication strategy for self-assembly. The molecule is designed with end groups (dull gold spheres) of sulfur atoms, which automatically assembly onto the gold wire contacts. The blue fuzz above and below the atoms represents the electron clouds, through which the current actually flows. Figure 4.10 is a representation of the experiments that demonstrated the first electrical measurement of a single atom (Reed et al. 1997).

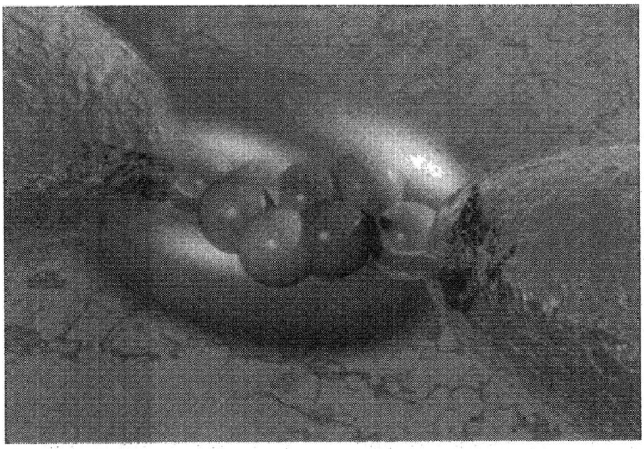

Figure 4.10. A single molecule bridging the gap between two metallic contacts, forming the smallest and ultimate limit of an electronic device (©1999 Mark Reed; all rights reserved).

4.7.10 Robotic Assembly of Nanostructures

Contact person: Ari Requicha, University of Southern California

Nanoparticles may be positioned accurately and reliably on a surface by using the tip of an atomic force microscope (AFM) as a robot. The AFM images the original, random distribution of particles in dynamic (non-contact) mode, and then pushes each particle along a desired trajectory by moving against the particle with the feedback turned off. Potential application of nanomanipulation to NanoCDs by using ASCII language has been illustrated. The new type of digital storage could have with densities several orders of magnitude larger than those of current compact disks. Nanomanipulation of nanoparticles with AFMs has been demonstrated at room temperature, in ambient air and in liquids. The resulting structures can be linked chemically, e.g., by using di-thiols or DNA as glue, to produce nano-components that can themselves be manipulated as sub-assemblies (Resch et al. 1998, Requicha, 1999). Robotic operations with standard, single-tip AFMs have low throughput, and are useful primarily for prototyping. Large-scale production requires massively parallel tip arrays, which are under development at several laboratories (see Section 3.2).

4.8 REFERENCES

Alivisatos, A.P., K.P. Johnsson, X.G. Peng, T.E. Wilson, C.J. Loweth, M.P. Bruchez, and P.G. Schultz. 1996. Organization of nanocrystal molecules using DNA. *Nature* 382:609-611.

Alivisatos, A.P., P.F. Barbara, A.W. Castleman, J. Chang, D.A. Dixon, M.L. Klein, G.L. McLendon, J.S. Miller, M.A. Ratner, P.J. Rossky, S.I. Stupp, and M.E. Thompson. 1998. From molecules to materials: Current trends and future directions. *Advanced Materials* 10:1297-1336.

Antonelli, D.M. and J.Y.Ying. 1996. Mesoporous materials. *Current Opinion in Colloid and Interface Science* 1:523-529.

Braun, E., Y. Eichen, U Sivan, and G. BenYoseph. 1998. DNA-templated assembly and electrode attachment of a conducting silver wire. *Nature* 391:775-778.

Bruchez, M., M. Moronne, P. Gin, S. Weiss, and A.P. Alivisatos. 1998. Semiconductor nanocrystals as fluorescent biological labels. *Science* 281:2013-2016.

Brus, L. 1996. Semiconductor colloids: Individual nanocrystals, opals and porous silicon. *Current Opinion in Colloid and Interface Science* 1:197-201.

Chan, W.C.W. and S.M. Nie. 1998. Quantum dot bioconjugates for ultrasensitive nonisotopic detection. *Science* 281:2016-2018.

Chen, J., M.A. Hamon, H. Hu, Y. Chen, A.M. Rao, P.C. Eklund, and R.C. Haddon. 1998. Solution properties of single-walled carbon nanotubes. *Science* 282: 95-98.

Chou, S.Y., P.R. Krauss, and P.J. Renstrom. 1996. Imprint lithography with 25-nanometer resolution *Science* 272:85.

Chou, S.Y., P.R. Krauss, W. Zhang, L. Guo and L. Zhuang. 1997. Sub-10 nm imprint lithography and applications. Invited, *J. Vac. Sci. Technol. B* 15(6):2897.

Chou, S.Y. and L. Zhuang. 1997. Unpublished.

Chou, S.Y. 1998. U.S. Patent No. 5,772,905.

Chou, S.Y. and L. Zhuang. 1999. Lithographically induced self-assembly of periodic polymer micropillar arrays. *J. Vac. Sci. Technol. B* 17(6):3197-3202.

Dai et al. 1996. Nanotubes as nanoprobes in scanning probe microscopy. *Nature* 384:147.

de Heer et al. 1995. A carbon nanotube field-emission electron source. *Science* 270:1179.

Guo et al. 1995. *Chem. Phys. Lett.* 243:49.

Kiehl, R.A., M Yamaguchi, O. Ueda, N. Horiguchi, and N. Yokoyama. 1996. Patterned self-assembly of one-dimensional arsenic particle arrays in GaAs by controlled precipitation. *Appl. Phys. Lett.* 68:478-480.

Liu et al. 1998. Fullerene pipes. *Science* 280:1253.

Martin, T.P. 1996. Shells of atoms. Physics Reports-Review section of *Physics Letters* 273:199-241.

Matthews, O. A., A.N. Shipway, and J.F. Stoddart. 1998. Dendrimers—branching out from curiosities into new technologies. *Progress in Polymer Science* 23:1-56.

McClelland, J., R.E. Scholten, E.C. Palm, and R.J. Celotta. 1993. Laser-focused atomic deposition. *Science* 262:877-880.

Mucic, R. C., J.J. Storhoff, C.A. Mirkin, and R.L. Letsinger. 1998. DNA-directed synthesis of binary nanoparticle network materials. *Journal of the American Chemical Society* 120:12674-12675.

Reed, M.A., C. Zhou, C.J. Muller, T.P. Burgin, and J. M. Tour. 1997. Conductance of a molecular junction. *Science* 278: 252-254.

Resch, R., C. Baur, A. Bugacov, B. E. Koel, A. Madhukar, and A. A. G. Requicha. 1998. Building and manipulating 3-D and linked 2-D structures of nanoparticles using scanning force microscopy. *Langmuir*, Vol. 14, No. 23, pp. 6613-6616, November 10.

Requicha, A.A.G. 1999. Nanoparticle patterns. *J. of Nanoparticle Research*. Vol. 1, No. 3, pp.321-323.

Rinzler et al. 1995. *Science* 269:1550.

Stupp, S.I., ed. 1998. Interdisciplinary macromelecular science and engineering. In *Proc. NSF Workshop*. U. of Illinois.

Stupp, S.I., V. LeBonheur, K. Walker, L.S. Li, K.E. Huggins, M. Keser, and A. Amstutz. 1997. Supramolecular materials: Self-organized nanostructures. *Science* 276:384-389.

Terrones, M., W.K. Hsu, H.W. Kroto, and D.R.M. Walton. 1999. *Nanotubes: A revolution in materials science and electronics*, Vol. 199, 189-234.

Thess et al. 1996. Crystalline ropes of metallic carbon nanotubes. *Science* 273:483.

Tomalia, D.A. 1994. Starburst cascade dendrimers—fundamental building-blocks for a new nanoscopic chemistry set. *Advanced Materials* 6:529-539.

Wong et al. 1998. Covalently functionalized nanotubes as nanometre-sized probes in chemistry and biology. *Nature* 394:52.

Xia, Y. et al. 1997. *Adv. Mater.* 9:147.

Chapter 5

APPLICATIONS: DISPERSIONS, COATINGS, AND OTHER LARGE SURFACE AREA STRUCTURES

Contact persons: **P. Wiltzius, Lucent Technologies; K. Klabunde, Kansas State University**

5.1 VISION

In the future, coatings will have improved properties due to nanoparticle incorporations and the methodology of incorporation. Coatings will also be ordered or patterned at the micro- and nano-levels. Similarly, dispersions, powders, and macroscopic bodies with unique morphologies and ordered structures will be discovered. These materials will revolutionize industries dealing with paints, corrosion protection, environmental remediation, drug delivery, printing and optical communications.

5.2 CURRENT SCIENTIFIC AND TECHNOLOGICAL ADVANCEMENTS

The area of dispersions and coatings has seen tremendous advances over the past decades. These advances cover the spectrum from scientific achievements resulting from long-term research to commercial successes. Several significant examples are detailed in the following paragraphs.

Nanostructured and Nanocomposite Films

A new field of scientific research has grown out of the new capability of creating monolayers of organic molecules on substrates (e.g., alkanethiols on gold) with well-organized crystalline order in the monolayer of the organic molecules (Dubois and Nuzzo 1992). A combination of self-assembly with new patterning tools such as microprinting and micromolding (see Xia et al. 1999 and references therein) provides new non-lithographic techniques for creating small patterns on planar and non-planar surfaces.

"Frictionless" films have been discovered over the past five years and promise great economic impact.

Nanostructured emitters like diamond (Zhu et al. 1998) and carbon nanotubes (Zhu et al. 1999) have been demonstrated to have far superior current characteristics compared to conventional field-emitting tips and will play an important role in displays.

Advanced scratch-resistant films of the future will have nanocomposite formulations. In addition, polymeric nanocomposites will enable tunable surface and bulk properties such as adhesion and barrier capabilities.

Organic Templates for Nanoscale Inorganic Synthesis

The discovery of the MCM-41 mesoporous silicate in 1989 at Mobil (Wise 1999) has created a new field in nanotechnology. This process uses liquid crystalline phases to synthesize silica with well-controlled pore sizes in the 1.5 to 10 nm range. Ten years after the discovery of this process, the first commercial products are appearing in the marketplace. Applications include catalysis, filtration, and separation. The same concept has been applied to the synthesis of other nanostructured materials, notably compound semiconductor nanocrystals (Brus et al. 1991) and mesoporous oxides (Yang et al. 1998). Optically transparent magnetic materials for storage applications have been designed using organic templates.

Nanocrystalline Powders and Consolidated Structures

Sol-gel and aerogel/hypercritical drying methods have allowed synthesis of inorganic oxides of huge surface areas with enhanced surface chemical adsorption properties. Under applied pressure these ultrafine powders can be consolidated into highly porous pellets with very large pore volumes and somewhat controllable pore size openings. The surface chemical properties of these ultrafine powders and consolidated pellets seem to be dependent on the unusual polyhedral shapes of the individual nanocrystals, and these materials have found applications as new superadsorbents for toxic chemicals and acid gases (Koper et al. 1997).

Three-Dimensional Assemblies of Nanoparticles

Colloidal gold coated with DNA strands has been used to assay the specific complimentary DNA (Mirkin et al. 1996; Mirkin et al. 1997; Alivisatos et al. 1996). Hybridization of the complementary DNA leads to aggregation of the colloid, accompanied by a color change.

Over the past year several groups have shown that self-assembled colloidal lattices (polystyrene or silica) can be used as scaffolds for inorganic replication. Either sol-gel chemistry or chemical vapor deposition (CVD) processes have prepared titania and carbon replicas (inverse structures) (Holland et al. 1998; Wijnhoven and Vos 1998). Likewise, nanocrystal superlattices (NCSL) of gold and cadmium sulfide/selenide have been formed that exhibit unique optical properties (Lin et al. 1999). These represent new forms of supramolecular crystalline matter. This opens great opportunities to manipulate light via management of void space.

Dispersions and Suspensions with Controlled Fluid Dynamic Properties

Recently, stable water-based ferrofluids have been prepared via organic templates. This has led to advances in colored magnetic inks for printers using magneto-hydrodynamic fluid management in such systems.

Dispersions and Suspension of Hydrophobic Materials

Because pharmaceutical compounds are often hydrophobic, suitable delivery vehicles must be sought (Gardner 1999). Nanoparticle formulations will often allow hydrophobic compounds to be assimilated directly by the body.

Also, as a means of extending imaging systems further, antihalation layers in photographic film systems now contain nanoparticulates to offer specific light filtration for imaging effects. In output media applications for digital printer systems, there are now nanoparticle inks that have been commercialized to achieve image permanency and improved color quality.

5.3 GOALS FOR THE NEXT 5-10 YEARS: BARRIERS AND SOLUTIONS

- Develop nanostructured field emitters (e.g., diamond, carbon nanotubes).
- Further develop the combination of colloidal self-assembly to create three-dimensional periodic structures (nanocrystal superlattices) and replication synthesis of high-dielectric materials (e.g., titania, compound semiconductors, metals, chalcogenide glasses) to produce new materials with interesting optical properties (e.g., photonic bandgap materials). These are likely to find applications in areas such as optical communications and laser technology.
- Further develop the synthesis of nanoscale ceramic materials from dispersions, creating ultrafine powders with new morphologies and composite materials with new properties (e.g., adsorptive, catalytic, thermal, structural, electrical, magnetic, and optical).
- Further utilize the synthesis of inorganic nanomaterials in nonaqueous media.
- If further developed, the next-generation mesoporous silicates and other oxides with highly controlled pore sizes (1 to several tens of nanometers) will play an important role as filters, catalysts, and structural frameworks for the generation of new materials.
- Utilize dispersions (microencapsulated inks, E-ink, Gyricon, Xerox, or polymer-dispersed liquid crystals) as components in electronic paper.
- Develop new ways of self-assembling nanoparticle colloids.
- Investigate synthesis of colloidal particles of controlled non-spherical morphology, (rods, dumbbells, pyramids, etc.). A better understanding of the template materials, such as random polymer matrices, nanotubes, and membranes, is necessary.
- Pursue sol-gel chemistry as a route to high quality silica, already entering the marketplace in the form of preforms for optical fiber

manufacturing. With further development, other areas where high quality silica is desired are likely to follow (see Section 5.7.1).

5.4 SCIENTIFIC AND TECHNOLOGICAL INFRASTRUCTURE

- In the area of coatings and dispersions, there is a significant need for better characterization tools, both ex-situ and in-situ (size distributions, morphologies, and interparticle forces). Current tools (light, X-ray, neutron scattering, microscopies, ultrasound spectroscopy, optical spectroscopy, etc.) all have severe limitations and give often model-dependent, average information.
- In this vein, scanning probe microscopies need to be developed to allow multispectral characterization of surfaces.
- Current characterization tools in the sub-optical regime are inherently two-dimensional (scanning electron microscopes, scanning probes, etc.). The discovery of characterization tools of three-dimensional structures at nanometer length scales is highly desirable (e.g., the equivalent of confocal light microscopy in the nanometer regime).
- Access to expensive equipment (transmission electron microscopes, lithography) and large facilities (synchrotrons, spallation sources, reactors, pilot plants for scaleup) should be enhanced, simplified, and facilitated (staffed appropriately).

5.5 R&D INVESTMENT AND IMPLEMENTATION STRATEGIES

- The field of nanotechnology is inherently multidisciplinary and will greatly benefit from strong interactions between universities, industry, and national labs. A general issue that needs to be solved is related to intellectual property: a task force might be able to work out guidelines, thus eliminating the reinvention of the wheel with each new interaction.
- In other fields (e.g., polymers), companies of all sizes exist that do custom synthesis. This capability is lacking in the area of nanoparticle synthesis. It is suggested that SBIRs be encouraged for nanoparticle synthesis, instrument development, and modeling.

5.6 PRIORITIES AND CONCLUSIONS

Nanocrystalline materials represent a bridge between molecules and solid state and exhibit properties that are unique. New coatings and dispersions will take advantage of the huge surface areas and the enhanced chemical reactivities at the surfaces of nanocrystalline materials. These surfaces can be modified with ligands, can be consolidated into porous solids, or can be incorporated into fluids or plastics. Technologies involving inks, magnetics, sorption, catalysis, optical coatings, drug delivery, paints, corrosion protection, and chemical/biological defense are or soon will be affected. However, there is a real need for improved environmentally acceptable synthetic methods for these unusual new materials.

Priorities for the future should emphasize chemical synthesis of nanoparticulate materials, on both the laboratory scale (grams) and pilot scale (kilograms). Academic, government, and industrial laboratories should be working together, and this suggests the need for enhanced funding for individual investigators who are collaborating with industry, and STTR/SBIR funding for academic/government/industry joint efforts. A second priority should be directed at improving methods of characterization, including extending the availability of new instrumentation, with appropriate infrastructure.

5.7 EXAMPLES OF CURRENT ACHIEVEMENTS AND PARADIGM SHIFTS

5.7.1 Optical Fiber Preforms Through Sol-Gel

Contact person: P. Wiltzius, Lucent Technologies

Optical fiber is made from preform cylinders, which are initially several inches in diameter. These glass cylinders are heated and drawn down into fiber with a diameter of typically 0.13 mm. The starting point of an optical fiber preform is commonly a synthetic quartz tube. The cost of these materials is substantial, and alternative manufacturing processes leading to high-quality glass are of great importance to the optical fiber industry.

Sol-gel processing of nanoparticles has emerged as a very promising route to low cost preforms. Typical processing steps include mixing of nanosized colloidal silica and additives, gelation, and casting into molds. A crucial step is the drying of the wet gel body without cracking. After purification (Figure 5.1), the body is consolidated to clear glass.

In addition to a dramatic cost reduction in fiber manufacture, there is also the promise of making novel glass compositions and fiber designs.

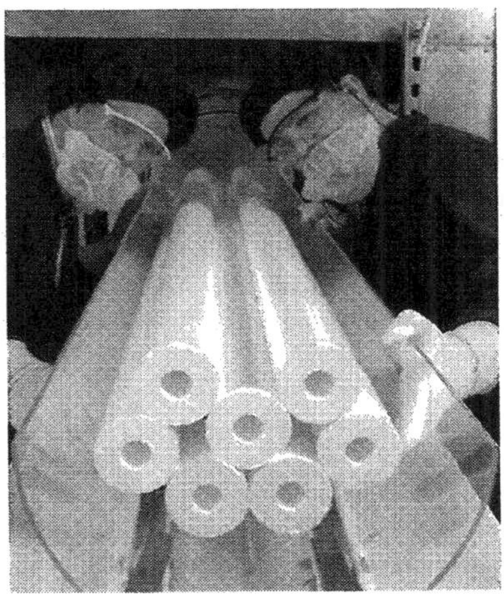

Figure 5.1. Photo of sol-gel preforms. The dried, unconsolidated tubes are being loaded into a silica "boat" to be heated in various gases to remove organic compounds, water, and refractory impurities (courtesy Bell Labs/Lucent Technologies).

5.7.2 Nanocomposites: Low-Cost, High-Strength Materials for Automotive Parts

Contact person: J. Garces, Dow Chemical Co.

Requirements for increased fuel economy in motor vehicles demand the use of new, lightweight materials, typically plastics that can replace metal. The best of these plastics are expensive and have not been adopted widely by U.S. vehicle manufacturers. Nanocomposites, a new class of materials under study internationally, consist of traditional polymers reinforced by nanometer-scale particles dispersed throughout (Figure 5.2). These reinforced polymers may present an economical solution to these problems. In theory, these new materials can be easily extruded or molded to near-final shape, yet provide stiffness and strength approaching that of metals—but at reduced weight. Corrosion resistance, noise dampening, parts consolidation, and recyclability all would be improved. However, producing nanocomposites requires development of methods for dispersing the particles throughout the plastic, as well as means to efficiently manufacture parts from such composites.

Dow Chemical Company and Magna International of America (in Troy, MI) have a joint Advanced Technology Program (ATP) project sponsored by the National Institute of Science and Technology (NIST) to develop practical synthesis and manufacturing technologies to enable the use of new high-performance, low-weight "nanocomposite" materials in automobiles (NIST 1997). Proposed potential applications would save 15 billion liters of gasoline and reduce carbon dioxide emissions by more than 5 billion kilograms over the life of one year's production of vehicles by the American automotive industry.

These materials are also likely to find use in non-automotive applications such as pipes and fittings for the building and construction industry; refrigerator liners; business, medical, and consumer equipment housings; recreational vehicles; and appliances.

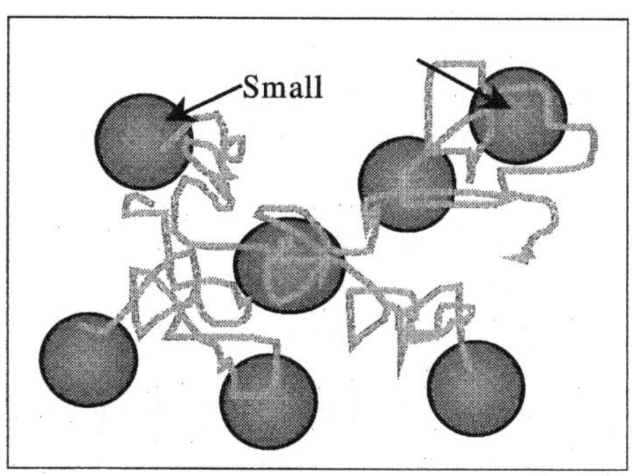

Figure 5.2. Schematic for nanoparticle-reinforced polymeric materials (after Schadler and Siegel 1998).

5.7.3 Biological Weapon Decontamination by Nanoparticles

Contact person: K. Klabunde, Kansas State University

Ultrafine powders can decontaminate mimics of biological weapons. For example, airborne heat-resistant *Bacillus Globigii* spores (a mimic for Anthrax) are killed at room temperature by airborne nanoparticles formulated from magnesium oxide and other reactive components. Figure 5.3 shows how the colony forming units (CFU) of *Bacillus Globigii* are detoxified over a 5-20 minute period. Similarly, *Bacillus Cereus* spores or *E. coli* bacteria can be disinfected. Several nanoparticle formulations have been shown to be effective, whereas commercial powders of the same materials are not.

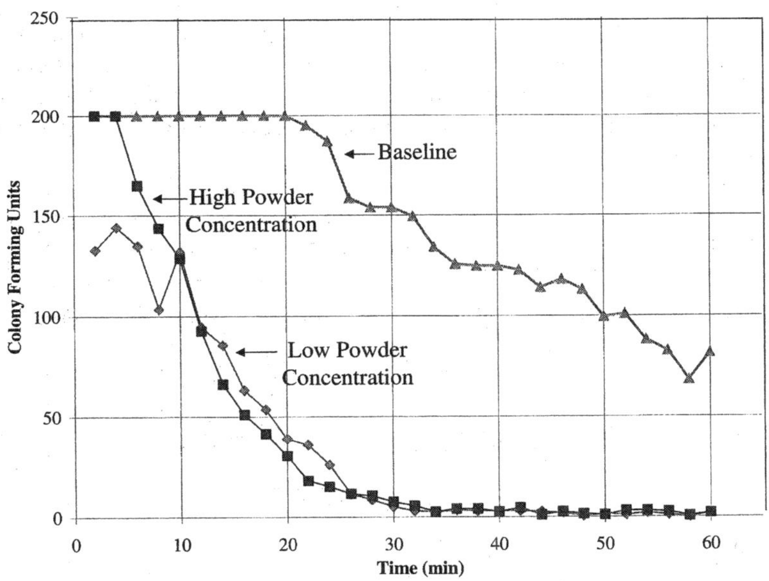

Figure 5.3. Decay curve for *Bacillus Globigii* baseline and *Bacillus Globigii* with chlorinated nanoparticle powder. It shows a baseline and two curves (each reproduced by two Brunswick air samplers), one for low nanoparticle concentration (10 mg/m^3) and one for a higher nanoparticle concentration (20 mg/m^3) in the air.

5.7.4 Nanoparticles for Use in Imaging Systems

Contact person: J. Mendel, Eastman Kodak Co.

Ability to generate thinner imaging layers is a consequence of using nanoparticulate preparations in image-recording layers such as graphic film systems; the high surface area of nanoparticulate filter dyes allows for higher feature performance and results in the ability to reduce the concentration of the dye component. In addition, the use of nanoparticulates in graphic film applications produces filter dye layers that have less light scatter by virtue of the finer particle size. This reduced light scatter leads to sharper absorption spectra, allowing for controlled handling of the film under specified light conditions.

Another advantage from nanoparticulate preparations of these filter dyes comes from the use of the polymeric media in the size reduction process. The polymeric media avoids creating attrition products, as are common with ceramic or metal media. The latter materials often produce pH changes, which adversely cause the dye to wander in the coating. Many attrition products affect dispersion stability and change the size reduction performance by influencing the final particle distribution of the slurry. Attrition products degrade milling performance and increase manufacturing maintenance costs. The polymeric media used in preparing nanoparticle slurries avoids those deficiencies, which results in a more invariant process for slurry preparation (U.S. Patents #5478705, 5442279, and 5474237).

5.7.5 Applications of Magnetic Fluids Containing Magnetic Nanoparticles

Contact person: T. Cader, Energy International, Inc.

Ferrofluids, first manufactured in their present-day form in the early 1960s, are colloids consisting of magnetic nanoparticles (10 nm in diameter) that are coated with a surfactant to ensure stability and suspended in a carrier such as transformer oil, water, or kerosene. The nanoparticles are individual permanent magnets, and when placed in suspension, the net magnetization of the ferrofluid is zero until a magnetic field is applied (see Figure 5.4). For example, a rotating magnetic field will align the nanoparticles, giving rise to a non-zero ferrofluid magnetization, while at the same time rotating the individual nanoparticles, as well as the ferrofluid itself, through entrainment of the carrier fluid. What distinguishes ferrofluids from other fluids are the body and surface forces and torques that arise in the ferrofluid when magnetic fields are applied to them, which in turn give rise to unusual fluid mechanical phenomena (Rosensweig 1985). Magnetic fluids such as magnetorheological suspensions employ larger particles (>100 nm), and unlike ferrofluids, tend to "freeze up" in intense magnetic fields.

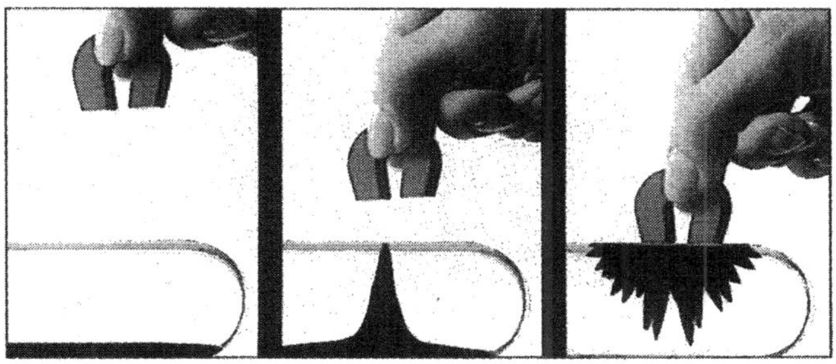

Figure 5.4. Visualization of a ferrofluid immersed in water (courtesy of Ferrofluidics Corp.).

Short-Term

Present-day applications require very small volumes of ferrofluids (~10 ml). This is one of the key reasons for current high prices of these fluids. Current applications, which have an estimated total market size of ~\$30-60 million, include the following:

- Contaminant exclusion seals on almost every PC disk drive manufactured, in silicon crystal-growing furnaces for the semiconductor industry, and in the medical field in MRI and CAT scan equipment
- Vacuum seals for high-speed, high-vacuum, motorized spindles
- Use as viscous dampers in the air gaps of stepper motors used in aircraft and various other machines

Medium-Term

Other applications are expected in three to five years. The success of key medium-term applications hinges in large part upon the advancement of the science of the behavior of ferrofluids subjected to time-varying and steady magnetic fields, for both heated and isothermal scenarios. For the time-varying fields, additional knowledge is necessary relating to the theory, including the appropriate boundary conditions for spinning nanoparticles at a wall. For all applications, good knowledge of the thermophysical properties (including dielectric properties) of the ferrofluids is essential. Medium-term applications currently under development include the following:

- Enhanced cooling and electrical insulation of power transformers
- Magnetic separation of ores in mining, and scrap metal separation

Success in any one of these high volume applications will lead to a significant reduction in ferrofluid pricing levels and will open the door to many more applications. For example, the potential market (in terms of revenue) for ferrofluid transformers lies in the range of $0.5-1 billion; in addition, there will be substantial savings to utilities.

Longer-Term

Applications for the longer term (over five years) are very promising. A host of exciting long-term applications are discussed in the literature. NASA's center for microgravity research is investigating how ferrofluids can play an important role in space, where gravity is essentially absent and free convection-dependent processes can be sustained by replacing the gravitational force with a magnetic force. In addition, through magnetic manipulation of the magnetic nanoparticles, ferrofluids offer a unique opportunity to remotely control the pressure, viscosity, electrical conductivity, thermal conductivity, and optical transmissivity of a fluid. Long-term applications also include the following:

- Development of ferrofluid-cooled and insulated power equipment for extended human space missions
- Development of nanoscale bearings that simultaneously levitate and lubricate a rotating shaft inside a bushing
- Magnetically controlled heat conductivity for precision temperature control of small devices such as electronic components

5.7.6 Summary of Current and Potential Applications

Contact person: J. Mendel, Eastman Kodak Co.

Table 5.1 summarizes current and potential future applications of nanostructured materials in the high surface area, coatings, and dispersions areas. More specific product applications are summarized below:

Commercial Applications Within the Next Three Years

- Use of nanoparticle metal oxides for decontamination of military warfare agents, which would be of great benefit to military personnel and useful for combating terrorism
- Use of nano, micro- and mesoporous composites (porous pellets) for air purification and disinfection, for example for airplanes and buildings
- Use of silica nanoparticles for manufacture of optical fibers

Commercial Applications Within the Next 5 Years

- Storage of information on transparent films or disks
- High quality magnetic inks
- Commercial firms offering custom nanoparticle syntheses
- Nanocrystalline superlative arrays used for the manufacture of new lasers
- A wide variety of new medicinal compounds
- Higher strength nanoparticle-polymer composites available to automobile and other industries
- More selective, efficient catalysts and sorbants and gas separation membranes

Table 5.1. Present and Future Applications of Nanostructured Dispersions

Now	3-5 years	Long term
Thermal barriers	Targeted drug delivery areas	Large fuel and energy advances from nanoparticles in fabrication and transportation
Optical (visible and UV) barriers	Gene therapy	Nanotechnology for improved environmental needs
Inkjet materials	Multifunctional nanocoatings	Nanoparticles for prosthesis and artificial limbs
Coated abrasive slurries	Nanocomposites for automobiles, weapon systems	Nanoparticles for integrated nanoscale sensors
Information recording layers	Nanocomposites for lighter, corrosion-resistant materials	Nanocomposites for space exploration uses
	Nanotechnology for taste enhancers, cosmetics, and other personal uses	Synthesis of nanomaterials in liquid nonaqueous media
		Next-generation mesoporous oxides
		Nanoparticles for decontamination

5.8 REFERENCES

Alivisatos, A.P., K.P. Johnsson, X. Peng, T.E. Wilson, C.J. Loweth, and P.G. Schultz. 1996. *Nature* 382:609.

Brus, L.E., M. Bawendi, W.L. Wilson, L. Rothberg, P.J. Carroll, T.M. Jedju, and M.L. Steigerwald. 1991. *Abstr. of Paper of the Amer. Chem. Soc.* 201:409-INOR, Part 1.

Drecker, S. and K.J. Klabunde. 1996. Enhancing effect of Fe_2O_3 on the ability of nanocrystalline calcium oxide to adsorb SO_2. *J. Am. Chem. Soc.* 118:12465-12466.

Dubois, L.H. and R.G. Nuzzo. 1992. *Annu. Rev. Phys. Chem.* 43:437-463.

Gardner, Colin. 1999. Statement on nanotechnology. In *IWGN Workshop Proceedings*, January 27-29, 1999, 346-53 (personal communication).

Holland, B.T., C.F. Blanford, and A. Stein. 1998. Synthesis of macroporous minerals with highly ordered three-dimensional arrays of spheroidal voids. *Science* 281:538-540.

Koper, O.B. and K.J. Klabunde. 1999 (in press). Development of reactive topical skin protectants against sulfur mustard and nerve agents. *Journal of Applied Toxicology.*

Koper, O., I. Lagadic, A. Volodin, K.J. Klabunde. 1997. Alkaline-earth oxide nanoparticles obtained by aerogel methods. Characterization and ration for unexpectedly high surface chemical reactivities. *Chem. of Materials* 9:2468-2480.

Lin, X.M., C.M. Sorensen, K.J. Klabunde. 1999. Ligand-induced gold nanocrystal superlattice formation in colloid solution. *Chem. of Materials* 11:197-202.

Mirkin, C.A., R.L. Lestinger, R.C. Mucic, and J.J. Storhoff. 1996. A DNA-based method for rationally assembling nanoparticles into macroscopic materials. *Nature* 382:607-609.

Mirkin, C.A. et al. 1997. Selective colorimetric detection of polynucleotides based on the distance-dependent optical properties of gold nanoparticles. *Science* 277:1078.

NIST (National Institute of Standards and Technology). 1997. ATP Project Brief 97-02-0047.

Rosensweig, R.E. 1985. *Ferrohydrodynamics.* New York: Cambridge University Press.

Schadler, L. and R.W. Siegel. 1998. Personal communication. Rensselaer Polytechnic Institute.

Wijnhoven, J.E.G.J., and W.L. Vos. 1998. Preparation of photonic crystals made of air spheres in titania. *Science* 281:802-804.

Wise, John J. 1999. A case history of commercialization of a breakthrough nanotechnology (MCM-41). In *IWGN Workshop Proceedings*, January 27-29, 1999, 284 (personal communication).

Yang, P.D., T. Deng, D.Y. Zhao, P.Y. Feng, D. Pine, B.F. Chmelka, G.M. Whitesides, and G.D. Stucky. 1998. Hierarchically ordered oxides. *Science* 282: 2244-2246.

Xia, Y., J.A. Rogers, K.E. Paul, G.M. Whitesides. 1999. Unconventional methods for fabricating and patterning nanostructures. *Chemical Reviews* 99(7):1823-1848.

Zhu, W., G.P. Kochanski, and S. Jin. 1998. Low-field electron emission from undoped nanostructured diamond. *Science* 282:1471-1473.

Zhu, W., C. Bower, O. Zhou, G. Kochanski, and S. Jin. 1999. Large current density from carbon nanotube field emitters. *Applied Physics Letters* 75(6):873-875.

Chapter 6

APPLICATIONS: NANODEVICES, NANOELECTRONICS, AND NANOSENSORS

Contact Persons: J. Jasinski, IBM; P. Petroff, University of California, Santa Barbara

6.1 VISION

In the broadest sense, nanodevices are the critical enablers that will allow mankind to exploit the ultimate technological capabilities of electronic, magnetic, mechanical, and biological systems. While the best examples of nanodevices at present are clearly associated with the information technology industry, the potential for such devices is much broader. Nanodevices will ultimately have an enormous impact on our ability to enhance energy conversion, control pollution, produce food, and improve human health and longevity.

6.2 CURRENT SCIENTIFIC AND TECHNOLOGICAL ADVANCEMENTS

Current Scientific Advances

In the past decade, our ability to manipulate matter from the top down, combined with advances and in some cases unexpected discoveries in the synthesis and assembly of nanometer-scale structures, has resulted in advances in a number of areas. Particularly striking examples include the following:

- The unexpected discovery and subsequently more controlled preparation of carbon nanotubes and the use of proximal probe and lithographic schemes to fabricate individual electronic devices from these materials (Iijimi 1991; Guo et al. 1995; Tans et al. 1997; Bockrath et al. 1997; Collins et al. 1997; Martel et al. 1998)

- The ability in only the last one or two years to begin to place carefully engineered individual molecules onto appropriate electrical contacts and measure transport through the molecules (Bumm et al. 1996; Reed et al. 1997)
- The explosion in the availability of proximal probe techniques and their use to manipulate matter and thereby fabricate nanostructures (Stroscio and Eigler 1991; Lyo and Avouris 1991; Jung et al. 1996; Cuberes et al. 1996; Resch et al. 1998)
- The development of chemical synthetic methods to prepare nanocrystals, and methods to further assemble these nanocrystals into a variety of larger organized structures (Murray et al. 1995)
- The introduction of biomolecules and supermolecular structures into the field of nanodevices (Mao et al. 1999)
- The isolation of biological motors, and their incorporation into nonbiological environments (Noji et al. 1997; Spudich et al. 1994)

Current Technological Advances

A number of examples of devices in the microelectronics and telecommunications industries rely on nanometer-scale phenomena for their operation. These devices are, in a sense, "one-dimensional" nanotechnologies, because they are micrometer-scale objects that have thin film layers with thicknesses in the nanometer range. These kinds of systems are widely referred to in the physics and electronics literature as two-dimensional systems, because they have two classical or "normal" dimensions and one quantum or nanoscale dimension. In this scheme, nanowires are referred to as one-dimensional objects and quantum dots as zero-dimensional. In this document, and at the risk of introducing some confusion, we have chosen to categorize nanodevices by their main feature nanodimensions rather than by their large-scale dimensions. Thus, two-dimensional systems such as two-dimensional electron gases and quantum wells in our notation are one-dimensional nanotechnologies, nanowires are two-dimensional nanotechnologies, and quantum dots are three-dimensional nanotechnologies. Examples include high electron mobility transistors, heterojunction bipolar transistors, resonant tunneling diodes, and quantum well optoelectronic devices such as lasers and detectors.

The most recent success story in this category is that of giant magnetoresistance (GMR) structures. These structures can act as extremely sensitive magnetic field sensors. GMR structures used for this purpose consist of layers of magnetic and nonmagnetic metal films. The critical layers in this structure have thicknesses in the nanometer range. The transport of spin-polarized electrons that occurs between the magnetic layers on the nanometer length scale is responsible for the ability of the structure to sense magnetic fields such as the magnetic bits stored on computer disks. GMR structures are currently revolutionizing the hard disk drive magnetic storage industry worth $30-40 billion/year (Prinz 1998; Disktrend 1998, Gurney and Grochowski 1998; Grochowski 1998). Our ability to control materials in one dimension to build nanometer-scale structures with atomic scale precision comes from a decade of basic and applied research on thin film growth, surfaces, and interfaces.

The extension from one nanodimension to two or three is not straightforward, but the payoffs can be enormous. Breakthroughs in attempting to produce three-dimensional nanodevices include the following:

- Demonstration of Coulomb blockade, quantum effect, and single electron memory and logic elements operating at room temperature (Guo et al. 1997; Leobandung et al. 1995; Matsumoto et al. 1996)
- Integration of scanning probe tips into sizeable arrays for lithographic and mechanical information storage applications (Lutwyche et al. 1998; Minne et al. 1996)
- Fabrication of photonic band-gap structures (Sievenpiper et al. 1998)
- Integration of nanoparticles into sensitive gas sensors (Dong et al. 1997)

6.3 GOALS FOR THE NEXT 5-10 YEARS: BARRIERS AND SOLUTIONS

In order to exploit nanometer-scale phenomena in devices, we must have a better understanding of the electronic, magnetic, and photonic interactions that occur on and are unique to this size scale. This will be achieved through experiment, theory, and modeling over the next decade. In addition, new methods to image and analyze devices and device components will be developed. These might include three-dimensional electron microscopies and improved atomic-scale spectroscopic techniques.

Over the same time period, we believe that it will become possible to integrate semiconductor, magnetic, and photonic nanodevices as well as molecular nanodevices into functional circuits and chips.

The techniques now being developed in biotechnology will merge with those from nanoelectronics and nanodevices. Nanodevices will have biological components. Biological systems will be probed, measured, and controlled efficiently with nanoelectronic devices and nanoprobes and sensors.

There will be significant progress in nanomechanical and nanobiomechanical systems, which will exhibit properties that are fundamentally different from their macroscopic counterparts.

There are important applications for instruments that will fly into space: nanocomponents are needed to achieve overall instrument sizes in the micron or millimeter range (http://www.ipt.arc.nasa.gov; http://www.cism.jpl.nasa.gov). Some of the same issues apply to battlefield sensors for situational awareness.

Finally, a significant goal is the development of nanometer-scale objects that manipulate and perform work on other nanometer-scale objects, efficiently and economically achieving the same things we currently rely on scanning tunneling microscopy (STM) or atomic force microscopy (AFM) to carry out. A first step towards this goal might be the integration of nanometer-scale control electronics onto micromachines.

Paradigm Shifts

In the information technology arena, nanodevices will both enable and require fundamentally new information processing architectures. Early examples of possible architectural paradigm shifts are quantum computation (Shor 1994; DiVincenzo 1995; Gershenfeld and Chuang 1997), quantum dot cellular automata (Lent and Tougaw 1997; Orlov et al. 1997), molecular electronics (Ellenbogen and Love 1999), and computation using DNA strands (Adleman 1994; Adleman 1998). Such architectures will fundamentally change the types of information technology problems that can be attacked. Effective implementation of these types of architecture will require nanodevices.

Other paradigm shifts include the emergence of quantized magnetic disks (Chou and Krauss 1996); single photonic systems (Kim et al. 1999) that will allow efficient optical communication; nanomechanical systems (Gimzewski et al. 1998); a broad class of structures and devices that merge biological and non-biological objects into interacting systems (Alivisatos et al. 1996; Mucic et al. 1998); and use of nanocomponents in the shrinking conventional circuit architectures (Ellenbogen and Love 1999).

Research on nanodevices using nanoscale wiring and molecular logic, as well as new principles for devices such as spin electronics, have made significant inroads in the past year or two.

6.4 SCIENTIFIC AND TECHNOLOGICAL INFRASTRUCTURE

The exploration and fabrication of nanodevices requires access to sophisticated and sometimes expensive tools. More and better access to such equipment as well as rapid prototyping facilities is needed. Of equal importance is the recognition that success in nanodevices will draw upon expertise from a broad range of traditional disciplines. Therefore, it is imperative that programs be established that facilitate and strengthen cross-fertilization among diverse disciplines and that allow rapid adoption of new methods across field boundaries.

6.5 R&D INVESTMENT AND IMPLEMENTATION STRATEGIES

Nanodevices are in some ways the most complicated nanotechnological systems. They require the understanding of fundamental phenomena, the synthesis of appropriate materials, the use of those materials to fabricate functioning devices, and the integration of these devices into working systems. For this reason, success will require a substantial funding level over a long period of time. There is strong sentiment for single investigator funding as well as for structured support of interdisciplinary teams.

6.6 CONCLUSIONS AND PRIORITIES

Priorities in Research and Development

- Development of new systems and architectures for given functions
- Study of interfaces and integration of nanostructures into devices and systems
- Multiscale, multiphenomena modeling and simulation of complex systems

Priorities in Modes of Support

- Establishment of consortia or centers of excellence for the research priorities identified above, by using vertical and multidisciplinary integration from basic research to prototype development
- Encouragement of system integration at the nanoscale in research and education

6.7 EXAMPLES OF CURRENT ACHIEVEMENTS AND PARADIGM SHIFTS

6.7.1 Organic Nanostructures: The Electrical Conductivity of a Single Molecule

Contact person: H. Goronkin, Motorola

By combining chemical self-assembly with a mechanical device that allows them to break a thin gold wire with nanometer scale control, researchers have succeeded in creating a "wire" consisting of a single molecule that can connect two gold leads (Figure 6.1). Using this structure, they have been able to begin to measure and study the electrical conductivity of a single molecule.

6.7.2 Molecular Electronics

Contact person: S. Williams, Hewlett-Packard

If the reduction in size of electronic devices continues at its present exponential pace, the size of entire devices will approach that of molecules within a few decades. However, well before this happens, both the physics upon which electronic devices are based and the manufacturing procedures used to produce them will have to change dramatically.

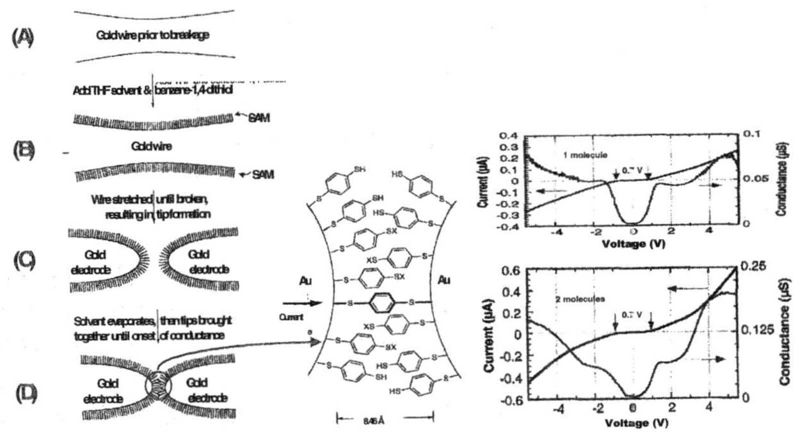

Figure 6.1. Organic nanostructures: on left, showing self-assembly of benzene-1,4-dithiol onto Au electrodes; on right, showing room-temperature I-V measurements suggesting presence of a Coulomb gap (reprinted with permission from Reed et al. 1997, ©1997 American Association for the Advancement of Science).

This is because current electronics are based primarily on classical mechanics, but at the scale of molecules, electrons are quantum mechanical objects. Also, the cost of building the factories for fabricating electronic devices, or fabs, is increasing at a rate that is much larger than the market for electronics; therefore, much less expensive manufacturing process will need to be invented.

Thus, an extremely important area of research is *molecular electronics*, for which molecules that are quantum electronic devices are designed and synthesized using the batch processes of chemistry and then assembled into useful circuits through the processes of self-organization and self-alignment. If molecular electronics achieves the ultimate goal of using individual molecules as switches and carbon nanotubes as the wires in circuits, we can anticipate nonvolatile memories with one million times the bit area density of today's DRAMs and power efficiency one billion times better than conventional CMOS circuitry. Such memories would be so large and power-efficient that they could change the way in which computation is performed from using processors to calculate on the fly to simply looking up the answer in huge tables.

A major limitation of any such process is that chemically fabricated and assembled systems will necessarily contain defective components and connections. This limitation was addressed in a 1998 paper entitled "A Defect-Tolerant Computer Architecture: Opportunities for Nanotechnology" (Heath et al. 1998). By describing a silicon-based computer that was designed to operate perfectly in the presence of huge numbers of manufacturing defects, researchers from Hewlett-Packard (HP) and the University of California–Los Angeles (UCLA) presented an architectural solution to the problem of defects in molecular electronics, as described in Figure 6.2, and thus demonstrated in principle that manufacture by chemical assembly is feasible.

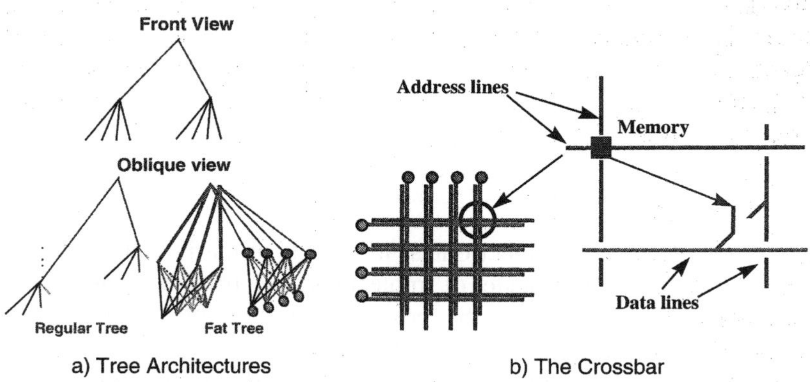

Figure 6.2. The logical design of a defect-tolerant circuit: (a) shows a "fat tree" architecture in which every member of a logical level of the tree hierarchy can communicate with every member at the next level; in the case of a defective component, this structure enables one to route around and avoid the defect; (b) shows how this architecture is implemented using cross bars, which are very regular structures and look like something that can be built chemically. The complexity required for a computer is programmed into the cross bars by setting the switches to connect certain elements of the tree together. Using silicon circuitry, two completely separate sets of wires (address and data lines) are required for the cross bars, and seven transistors are required for each switch, since a continual application of electrical power is required to hold the sense of the switches.

In 1999, researchers from HP Labs and UCLA experimentally demonstrated the most crucial aspect for such a system, an electronically addressable molecular switch that operates in a totally "dry" environment (Collier et al. 1999). As illustrated in Figure 6.3, logic gates were fabricated from an array of configurable molecular switches, each consisting of a monolayer of electrochemically active rotaxane molecules sandwiched between metal electrodes.

Figure 6.4 illustrates the operation of the switches. In the "closed" state, current flow is dominated by resonant tunneling through the electronic states of the molecules. The switches are irreversibly opened by applying an oxidizing voltage across the device. In this case, since the memory of the molecules is not volatile, only one set of wires is needed to set and read the state of the molecules, and in principle, one molecule can replace seven transistors in a conventional silicon circuit. In the demonstration, several devices were configured together to produce AND and OR logic gates. The high/low current levels of those gates were separated by factors of 15 and 30, respectively, which is a significant enhancement over that for conventional wired-logic gates.

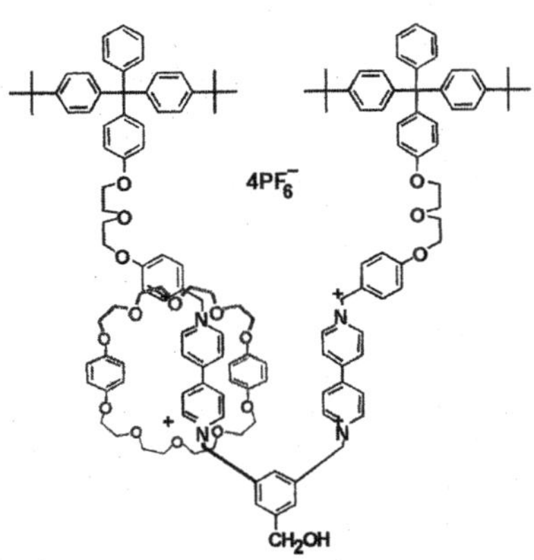

Figure 6.3. The atomic structure of one of the molecular switches used in the devices described above, which is known as a rotaxane (F. Stoddart, UCLA). This molecule conducts via resonant tunneling through unoccupied molecular orbitals when it is in its reduced chemical state (switch closed), but it is a tunneling barrier in its oxidized state (switch open). The switch can be closed electronically in a solid-state circuit by applying the appropriate voltage across the molecule (Balzani et al. 1998; Credi et al. 1997).

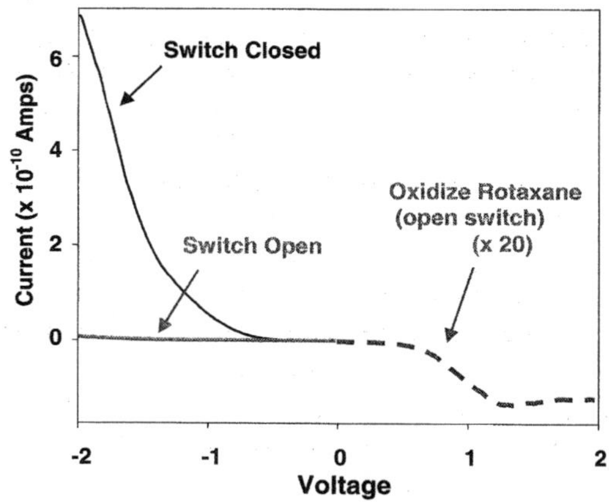

Figure 6.4. The current-voltage (I-V) characteristic of a large number of molecular switches is shown in both the "on" and "off" states. Initially, the molecular switches are closed, and applying a negative voltage across the molecules results in a "large" current flow that varies exponentially with the magnitude of the applied voltage. This portion of the I-V curve is highly reproducible until the potential across the molecule exceeds +1 V. This voltage irreversibly oxidizes the switches, and after this process, applying a negative voltage results repeatedly in a "small" current, demonstrating that the switch is open.

6.7.3 Molecular Logic

Contact persons: J.M. Tour, Rice University, and M. Reed, Yale University

Electron transport studies in molecular-scale systems have recently become possible with the utilization of advanced microfabrication and self-assembly techniques (Aviram and Ratner 1998; Petty et al. 1995). Investigations are now possible of the electronic conduction through conjugated molecules that are end-bound onto surfaces; these have been demonstrated with a scanning tunneling microscope (Bumm et al. 1996), with micromachined silicon nanopores (Zhou et al. 1997), and with proximal probes (Reed et al. 1997; Kergueris 1999). Work on the proximal probes demonstrated that 0.1 microamp of current can be transported through a single molecule (Reed et al. 1997). However, in all of the past embodiments, the electronic properties exhibit simple diodic behavior that is unsuitable for potential circuit applications. Researchers recently observed the first large and useful reversible switching behavior in an electronic device that utilizes molecules as the active component. That work is disclosed here (Chen et al. 1999).

The essential feature of the fabrication process is the employment of nanoscale device area that gives rise to a small number of self-assembled molecules (ca. 1,000), which eliminates pinhole and other defect mechanisms that hamper through-monolayer electronic transport measurements. This technique has demonstrated good control over the device area and intrinsic contact stability and produces a large number of devices with acceptable yield so that statistically significant results can be produced (Figure 6.5).

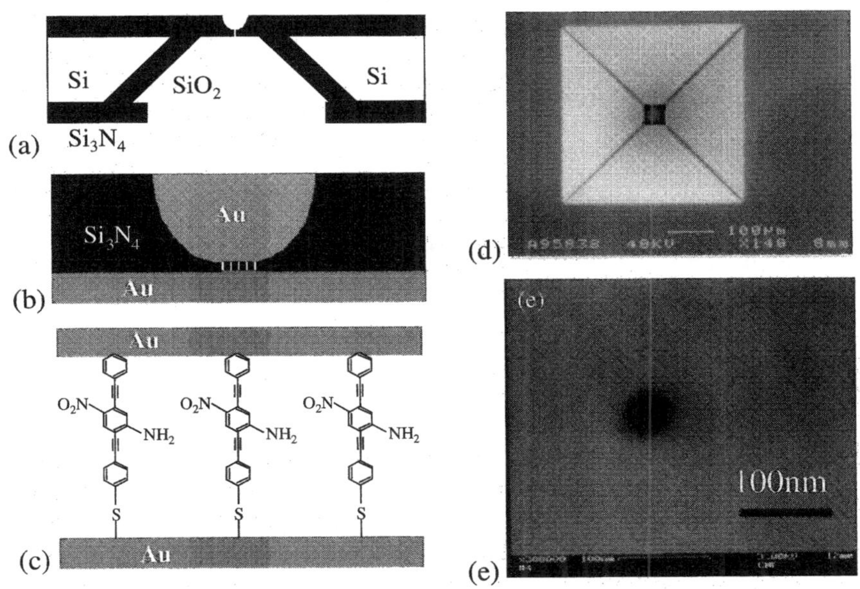

Figure 6.5. Schematics of device fabrication: (a) cross section of a silicon wafer with a nanopore etched through a suspended silicon nitride membrane; (b) Au-SAM-Au junction in the pore area; (c) blowup of the active SAM region with compound **1c** sandwiched in the junction; (d) SEM micrograph of pyramid Si structure after unisotropic Si etching, i.e., the bottom view of (a); (e) SEM micrograph of etched nanopore through silicon nitride membrane (reprinted with permission from Chen et al. 1999, ©1999 American Association for the Advancement of Science).

The active electronic component (synthesis shown in Figure 6.6) was made from an organic compound upon exposure to Au. Figure 6.7 illustrates the I-V characteristics of the Au-(**1c**)-Au devices at 60 K. The I-V is fully reversible upon change in bias sweep direction. This is the first observation of robust and large negative differential resistance (NDR) in a device where molecules form the active region with peak-to-valley-ratios (PVRs); and the PVRs here are >1000:1. The performance exceeds that observed in typical solid state quantum well resonant tunneling heterostructures. Therefore, in addition to the obvious size advantages for scaling, the intrinsic device characteristics (i.e., valley current shutoff) may be superior to solid state embodiments; present silicon devices rarely exceed PVRs of 100:1.

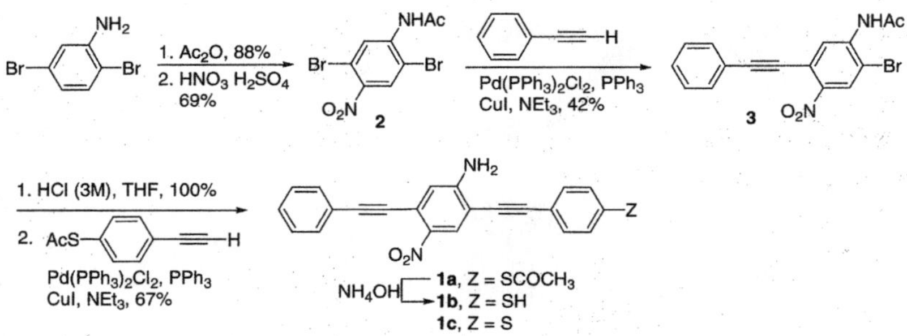

Figure 6.6. Schematic of the synthesis of the active molecular compound and its precursors (**1a-c**).

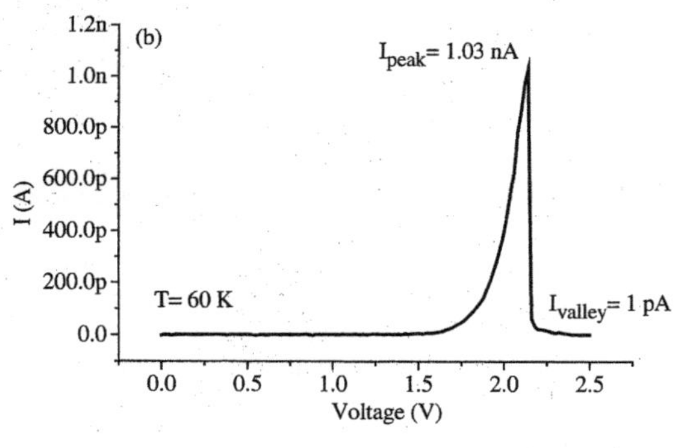

Figure 6.7. I-V characteristics of the Au-(2'-amino-4-ethynylphenyl-4'-ethynylphenyl-5'-nitro-1-benzenethiolate)-Au devices at 60 K. The peak current density is ~50 A/cm^2, the NDR is ~ -400 $\mu\Omega$-cm^2, and the PVR is 1030:1 (reprinted with permission from Chen et al. 1999, ©1999 American Association for the Advancement of Science).

6.7.4 A Field-Effect Transistor Made from a Single-Wall Carbon Nanotube

Contact person: P. Avouris, IBM Research

Several research groups around the world have succeeded in fabricating electrical switches such as the field-effect transistor from single-walled carbon nanotubes. In the case illustrated in Figure 6.8, a single-walled carbon nanotube 1.6 nm in diameter was manipulated into place using an atomic force microscope. Once placed on the metal contacts, the semiconducting tube behaved like the channel in a field-effect transistor, turning on or off depending on the applied gate voltage. Nanotubes hold great promise as electronic elements for a variety of different nanostructures. Researchers are just beginning to understand how they conduct electricity and how to place them into appropriate device structures. It is interesting to note that both the atomic force microscope used to fabricate this structure and the carbon nanotubes that form the critical element were developed only in the past decade (Martel et al. 1998).

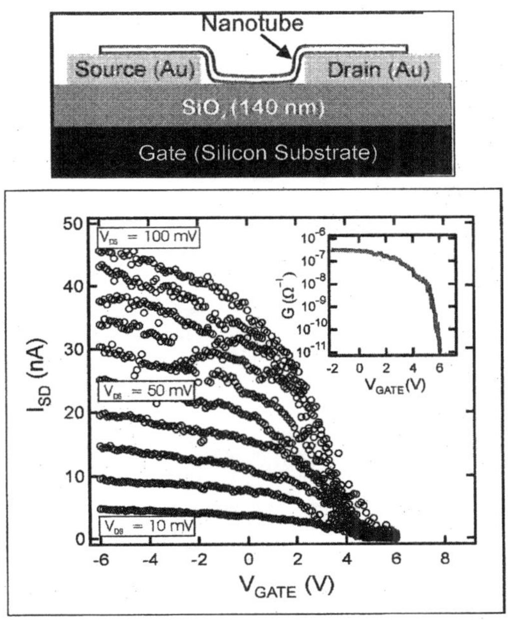

Figure 6.8. Field-effect transistor based on a single 1.6 nm diameter carbon nanotube (adapted from Martel et al. 1998, reprinted by permission; ©1998 American Institute of Physics).

6.7.5 A Commercial IBM Giant Magnetoresistance Read Head

Contact person: E. Grochowski, IBM

When certain kinds of materials systems are exposed to a magnetic field, their electrical resistance changes. This effect, called the magnetoresistive effect, is useful for sensing magnetic fields such as those in the magnetic bits of data stored on a computer hard drive. In 1988, the giant magnetoresistance effect was discovered in specially prepared layers of nanometer-thick magnetic and nonmagnetic films. By 1991, work at the IBM Almaden Research Center demonstrated that the GMR effect could be observed in easily made samples and that a special kind of GMR structure, a spin valve, could sense very small magnetic fields. This opened the door to the use of GMR in the read heads for magnetic disk drives. IBM first announced a commercial product based on this design in December 1997. In the spin valve GMR head shown in Figure 6.9, the copper spacer layer is about 2 nm thick, and the cobalt GMR pinned layer is about 2.5 nm thick. The thickness of these layers must be controlled with atomic precision.

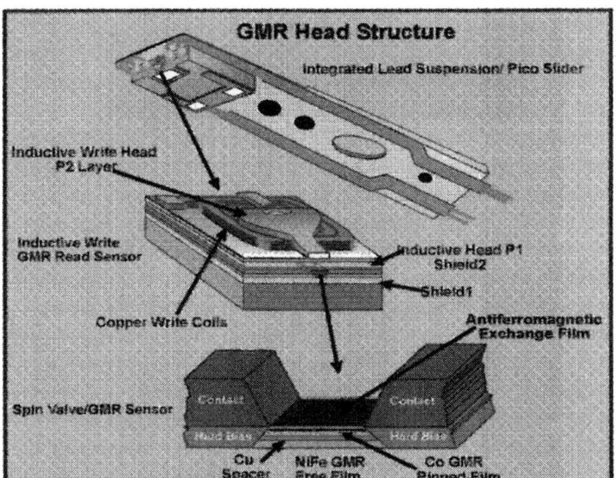

Figure 6.9. Commercial IBM giant magnetoresistance read head.

6.7.6 Nanoelectronic Devices

Contact person: G. Pomrenke, Defense Advanced Research Projects Agency (on detail from Air Force Office of Scientific Research)

Nanoelectronics offers a broad set of opportunities by focusing on quantum devices and addressing their potential for high performance through increases in density (factors of 5 to 100), speed (factors of 10 to 100), and reduced power (factors of more than 50) (see Figure 6.10). Resonant tunneling devices are being explored with demonstrated successes in multivalued logic and various logic circuits and memory circuits. SET logic and memory concepts are being explored with focus on memory applications. Molecular electronics and self-assembly approaches have shown a path towards manufacturing alternatives and device options for regimes beyond traditional scaling. Spin devices in the form of nanomagnetics using the magnetoresistive effect in magnetic multilayers have demonstrated their use for nonvolatile, radiation-hard memory. Quantum cellular automata and coupled quantum dot technology are being explored and their potential assessed for transistorless computing. By exploring Si-based heterojunctions, bandgap engineering, vertical device structures, and quantum devices, inroads are being made into extending CMOS capabilities. Potential applications are in digital radar, electronic support measures (ESM) receivers, ATM data stream processing, wide bandwidth communications, digital image processing, waveform generation, and the broad area of analog to digital (A/D) applications. Demonstrations have shown the efficacy of resonant tunneling devices in various network environments. The long-term vision for nanoelectronics sees the use of quantum devices in other high performance systems especially in telecommunications for signal processors and electronics for A/D converters in detectors.

6.7.7 Resonant Tunneling Devices in Nanoelectronics

Contact person: G. Pomrenke, Defense Advanced Research Projects Agency (on detail from Air Force Office of Scientific Research)

Resonant tunneling and other tunneling devices have had a history spanning almost three decades; however, it was not until 1997 that these devices could be seriously considered as part of functional circuits. The crucial technology for advancing these quantum devices has been epitaxial growth and process control at the nanoscale. This has meant control at the atomic layer level, resulting in flexible manufacturing, long-term process repeatability, and first-pass success. The resonant tunneling diode (RTD) consists of an emitter and collector region, and a double-tunnel barrier structure that contains a quantum well, as shown in the energy band diagrams of Figure 6.11. This quantum well is so narrow (5-10 nm) that it can only contain a single so-called "resonant" energy level.

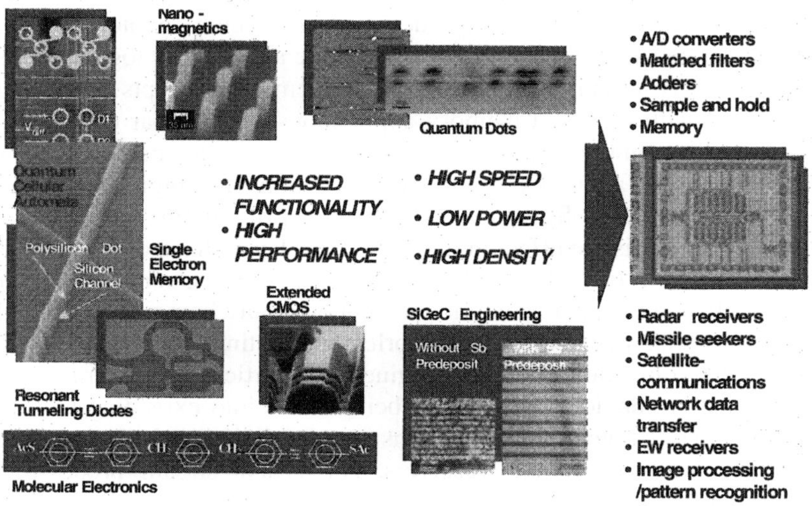

Figure 6.10. Nanoelectronics: device and architecture options for high-performance electronics.

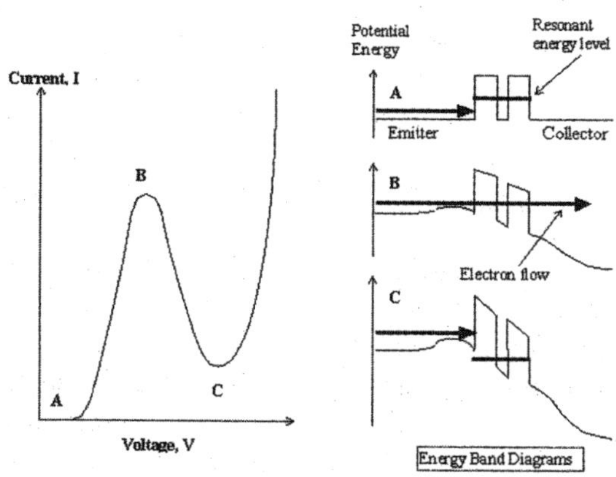

Figure 6.11. Resonant tunneling device (Moffat 1999).

The principle of this device is that electrons can travel from the emitter to the collector only if they are lined up with this resonant energy level. Initially, with a low voltage across the device (at point A), the electrons are below the point of resonance, and no current can flow through the device. As the voltage increases, the emitter region is warped upwards, and the collector region is warped downwards. Eventually, the band of electrons in the emitter line up with the resonant energy state and are free to tunnel through to the right. This gives an increase in the current up to the peak at point B. As the voltage across the device increases, the electrons are pushed up past the resonant energy level and are unable to continue tunneling. This can be observed by the drop in current to the valley at point C. As the voltage continues to increase, more and more electrons are able to flow over the top of the tunnel barriers, and the current flow rises. The current-voltage characteristic of this device is similar to that of the Esaki tunnel diode, in that it exhibits a peak and a valley in the curve. The difference is that RTDs have a much lower device capacitance, which allows them to oscillate faster, and their current-voltage characteristics (i.e., the positions of the peak and the valley) can be shaped with the appropriate bandgap engineering.

DARPA's Ultra Electronics Program accomplished the invention and simulation of a compact adder circuit with GHz speeds using redundant digit, multivalued logic, and the world's first demonstration of an integration process for yielding the core circuit elements needed for adders (see Figure 6.12), signal processors, and multivalued logic circuits. The technology developed was subsequently transferred into circuit development efforts, which have led to the demonstration of a 4 bit 2 GHz analog-to-digital converter, 3 GHz (40 dB spur-free dynamic range) clocked quantizer, 3 GHz sample and hold (55 dB linearity), clock circuits, shift registers, and ultralow power SRAM (50 nW/bit) (Seabaugh 1998). The "invention" of functional devices based on quantum confinement occurred in the early 1980s. In the optoelectronic area a good example is the self-electro-optic effect device (SEED), based on the quantum-confined Stark effect, for photonic switching applications. Another example is the vertical cavity surface-emitting laser (VCSEL), the backbone of optical communications. The technology offers two-fold speed increases, almost 10 times lower component counts, and 10 to 2,000 times lower power over conventional devices.

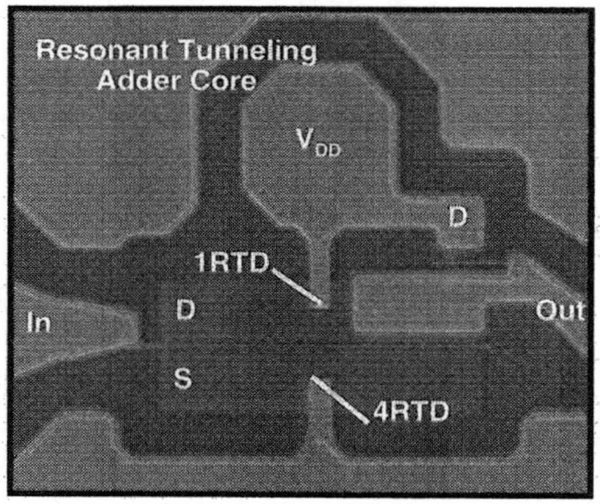

Figure 6.12. Resonant tunneling adder core (Seabaugh 1998).

6.7.8 Nanodevices and Breakthroughs in Space Exploration

Contact person: N.B. Toomarian, Jet Propulsion Laboratory

After more than three decades of exploring space, the National Aeronautics and Space Administration (NASA) has completed an initial reconnaissance of our solar system. The next missions will involve sending spacecraft to destinations that are much more difficult to travel to, like the Sun or Pluto. Also, spacecraft will be required to perform more difficult tasks, such as landing on a celestial body, collecting a sample of its material, and returning it to Earth. To carry out such technically challenging missions at an affordable cost, NASA has created the Deep Space Systems Technology Program, known as X2000. Every two to three years starting in the year 2000, the program will develop and deliver advanced spacecraft systems and body structures to missions bound for different areas of the solar system and beyond. In order to achieve reduction in the size of spacecraft, the avionics systems of the spacecraft are being reduced in size with each delivery of X2000, in part by means of integrating nanotechnology with microtechnology. Figure 6.13 attempts to chart the forecasts of the mass, volume, and power of future avionics systems of spacecraft. The leftmost column shows the Mars Pathfinder spacecraft, which represents the current state of the art.

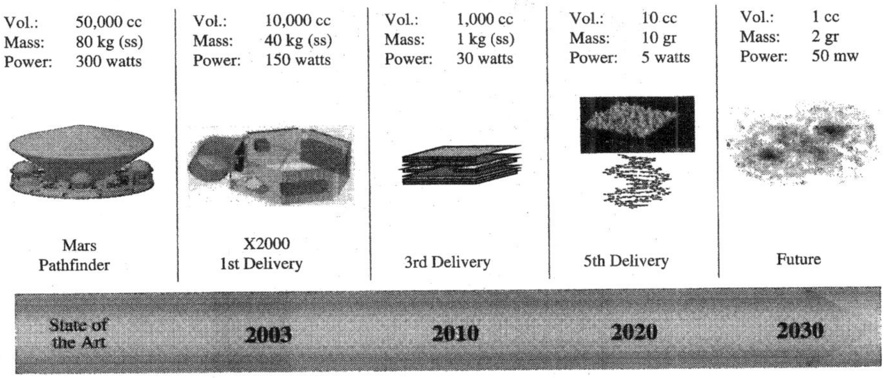

Figure 6.13. Avionics roadmap.

The first delivery for X2000 is an integrated avionics system that subsumes the functionalities of command and data handling, attitude control, power management and distribution, and science payload interface. Advanced packaging technologies as well as advanced design automation techniques are used to define a highly integrated, modular, building-block architecture for highly reliable and long-term survivable deep-space planetary missions. Advanced low-power techniques and architectures will drastically reduce overall power consumption compared to currently available flight hardware.

"System On A Chip" (SOAC) will prototype single-chip and multichip module solutions that lead towards an avionics system on a chip. This chip will integrate the avionics system that is being developed for the X2000 avionics deliverable. That is, the chip will include power management, sensor technology, and telecommunications modules, together with CPU and storage technology. To accomplish this, nanotechnology will be needed to miniaturize and integrate the different subsystems.

The goal for the year 2020 is to establish and maintain a world-class program to research revolutionary computing technologies (RCT) that will not only take us beyond the limits of semiconductor technology scaling but also will enable the vision of a "thinking spacecraft." A thinking spacecraft would be a totally autonomous, highly integrated, extremely capable spacecraft that operates at ultralow power. To achieve this goal, without a doubt, we need to employ nanoscience. In spite of the phenomenal advances in digital computing in recent years and those expected in the near future, even future supercomputers cannot compete with biological systems in performing certain ill-defined tasks such as pattern recognition, sensor fusion, fault-tolerant control, and adaptation to the environment. Biological systems address these types of problems with extreme ease and very low power. The forth column from the left in Figure 6.10 (Fifth Delivery) depicts two different technologies based on nanoscience that may have a great impact on the capabilities of our spacecraft by the year 2020:

- *Quantum computing*, that is, a joint venture between computer science and quantum physics. Although, the concept of a quantum computer is simple, its realization is not. Two issues motivate quantum computing:
    - Quantum mechanical concepts must be applied to solve intractable (NP-complete) computing problems.
    - From a computer miniaturization point of view, the size limit of a bit of information is important. Recently, this issue has attracted increased attention, due to the current development of nanotechnology and the design problems of semiconductor and metal devices that are approaching the quantum size limit. Consequently, the idea of quantum computing, in which the elements that carry the information are atoms, has attracted the attention of many scientists.
- *Biomimetics*, that is, systems or technologies inspired by architectures, functions, mechanisms, and principles found in biological systems, for example:
    - One gram of DNA could possibly store all the data in the Library of Congress.
    - The human brain contains about 10^{14} interconnects and operates at 10^{16} operations per second, using ultra low power and imprecise computing elements.
    - Humans are endowed with an immune system that provides recovery from illness—a "self-repair system."

As devices become smaller, lighter, and consume less power, NASA will be able to design and fly space probes on missions that are not currently possible.

6.7.9 A Biological Nanodevice for Drug Delivery

Contact person: S. Lee, Monsanto Corporation

The nanobiological anticancer agent PK1 (Figure 6.14) exploits the enhanced permeability and retention (EPR) effect associated with disease tissues with low integrity vasculature in order to deliver cytotoxin (doxorubicin) to tumors. The synthetic backbone of PKI (N-2-(hydroxypropyl) methacrylimide or HPMA) gives the complex a size (diameter in the mid-nanometer range) that makes it unable to extravasate efficiently into healthy tissues with normal vasculature. Tumor vasculature is abnormally permeable, allowing preferential accumulation of PK1 in tumor tissue. HPMA-bound doxorubicin is non-toxic, limiting toxicity to healthy tissue, and active doxorubicin is released from the complex preferentially in tumor tissues. The labile peptidic linker tethering doxorubicin to HPMA was selected because it is the substrate for a protease known to be over-expressed in the target tumor types. PK1 increases the tolerated doxorubicin dose by more than an order of magnitude by virtue of EPR-based targeting and its engineered tumor-preferred doxorubicin release properties. It is in human clinical trial in Europe. Contemplated embellishments to this and similar polymer therapeutics include use of monodisperse nanopolymers (dendritic polymers) to enhance control of EPR properties, incorporation of protein docking domains that recognize tumor associated antigens to tether the complex following its delivery to the tumor, and incorporation of additional antitumor agents thought to have synergistic effects with cytotoxins, that is, angiostatic agents, among others (Duncan 1997).

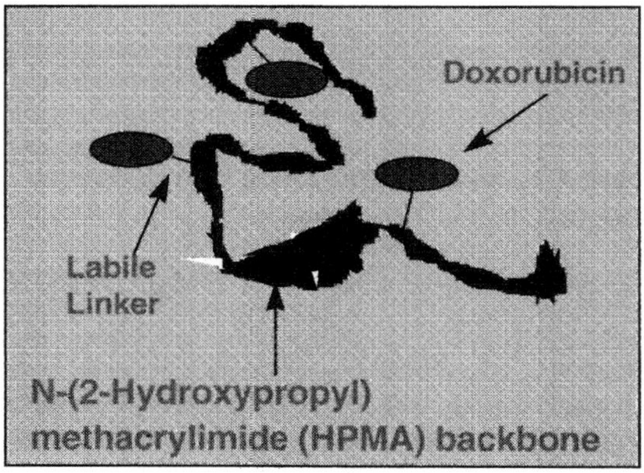

Figure 6.14. The nanobiological anticancer agent PK1 (Lee 1998).

6.7.10 Nanotechnology on a Chip: A New Paradigm for Total Chemical Analysis Systems

Contact person: T.A. Michalske, Sandia National Laboratories

The ability to make chemical and biological information much cheaper and easier to obtain is expected to fundamentally change healthcare, food safety, law enforcement, national security, and many other areas of direct interest to the American public. The vision of broadly available chemical analysis is fueling an international effort to develop "μChemLab™-on-a-chip" technology. Micro-total analysis systems (μ-TAS, as they are often referred to) are distinguished from simple sensors because they conduct a complete analysis; a raw mixture of chemicals goes in and an answer comes out. Sandia National Laboratories is developing a hand-held μChemLab™ demonstrator that will analyze for air-born chemical warfare agents and liquid-based explosives agents. The μChemLab™ development project brings together an interdisciplinary team of about 50 staff members from throughout the laboratory in areas of expertise including microfabrication, chemical sensing, microfluidics, and information sciences. Although nanotechnology plays an important role in current μTAS efforts, most μTAS approaches use miniaturized versions of conventional architecture and components to achieve system tasks. Small valves, pipes, pumps, separation columns, etc. are patterned after their macroscopic counterparts. Even though we are finding that these miniaturized components can work as well as (and sometimes better than) their macroscopic analogs, they simply will not allow for the vision of chemical laboratories in a grain of sand.

Nanotechnology will enable a completely new architecture, or nano-TAS. The ability to build materials with switchable molecular functions could provide completely new approaches to valves, pumps, chemical separations, and detection. For example, fluid streams could be directed by controlling surface energy without the need for a predetermined architecture of physical channels. Switchable molecular membranes and the like could replace mechanical valves. By eliminating the need for complex fluidic networks and micro-scale components used in current μ-TAS efforts, nano-TAS is a fundamentally new approach to allow greater function in much smaller, lower power total chemical analysis systems.

6.7.11 The Development of Useful Nanotech Robotic Systems

Contact person: M.W. Tilden and T.C. Lowe, Los Alamos National Laboratory

It is potentially feasible to manufacture nano-robots that are capable of sophisticated symmetric behaviors, either through independent function or by assembling themselves into collective units. Imagine, for example, high-resolution video screens that can repair themselves simply by having a microscopic robot at each screen element. These "pixelbots" would be capable of producing light, but also be smart enough to remove themselves from the video array should they ever fail. Other pixelbots would sense the vacancy left by any defective device and reorganize themselves to fill the hole. Another example is the incorporation of autonomous "nurse" robots into the human body that are chemically benign but are capable enough to remove cancer cells at the source. Having the ability to discriminate between healthy and cancerous cells, the "nurse-bots" would function independently to continuously heal tissue in ways beyond the current ability of the human body. The same notion can be implemented in self-optimizing silicon memories or processors, where a blown transistor would mean one just had to wait for the computer to heal itself.

One key to establishing such capabilities is research into autonomous self-assembled systems. Researchers at Los Alamos National Laboratory are exploring these systems by creating very inexpensive macro-scale robots. These sense and adapt to their environment, including assimilating other robots to execute such tasks as searching for and marking the location of unexploded land mines (Figure 6.15). The capabilities of these intelligent cellular systems are readily scaled, providing untapped possibilities for large numbers of inexpensive nano-machines to become microscopic building blocks for heretofore unimaginable functions—a form of "nano-Lego" for the new millennium with novel, untapped market potential.

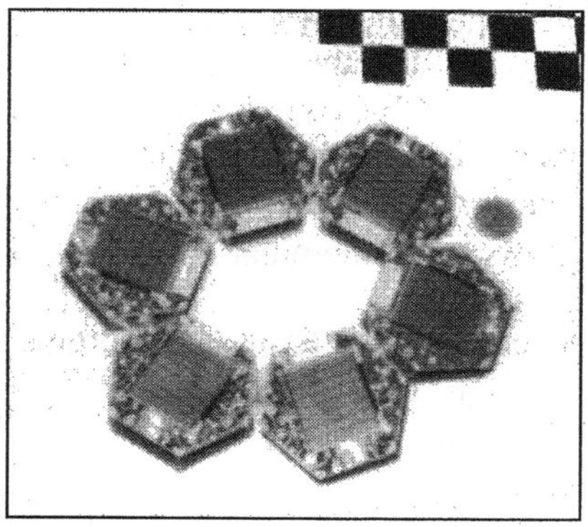

Figure 6.15. Models for nanoscale: Three-inch-diameter self-assembled robots mark the spot where an unexploded mine rests under the surface. Such robots are cheap, solar-powered, and have no processor to make application or miniaturization difficult.

6.7.12 Integrated Nanotechnology in Microsystems

Contact person: S.T. Picraux, Sandia National Laboratories

Advances in nanotechnology will have a profound effect on the future of integrated microsystems. The integration of microelectronic, microelectromechanical, optical, and chemical microsensors into "systems on a chip" is an area that may involve mechanical, optical, and/or chemical functions as well. As illustrated in Figure 6.16, these advances will make possible miniaturized systems that sense, think, talk (communicate), and act. However, these microscale systems will only become a reality if enabled by the control of performance at the nanoscale. Thus, for example, advances in micro-electromechanical systems (MEMS) and photonics shown in the figure depend on discoveries in nanoscience and nanoscale fabrication.

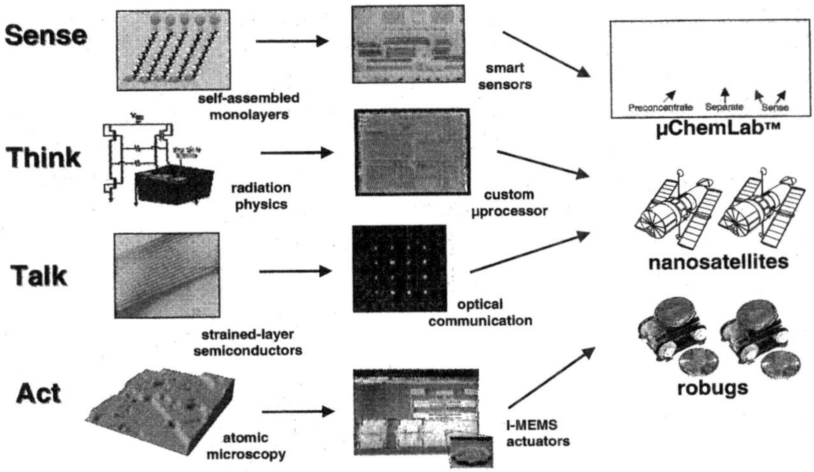

Figure 6.16. The control of mechanical, electrical, optical, and chemical properties at the nanoscale will enable significant improvements in integrated microsystems.

6.8 REFERENCES

Adleman, L. 1994. Molecular computation of solutions to combinatorial problems. *Science* 266:1021.

_____. 1998. Computing with DNA. *Scientific American* 279:34.

Alivisatos, A.P., et al. 1996. Organization of 'nanocrystal molecules' using DNA. *Nature* 382:609.

Aviram, A., and M. Ratner, eds. 1998. *Molecular electronics: Science and technology*. Annals of the New York Academy of Sciences, Vol. 852. New York: New York Academy of Sciences.

Balzani, V., M. Gomez-Lopez, and J.F. Stoddart. 1998. Molecular machines. *Acc. Chem. Res.* 31:405.

Bockrath, M., et al. 1997. Single electron transport in ropes of carbon nanotubes. *Science* 275:1922.

Bumm, L.A., J.J. Arnold, M.T. Cygan, T.D. Dunbar, T.P. Burgin, L. Jones II, D.L. Allara, J.M. Tour, and P.S. Weiss. 1996. Are single molecular wires conducting? *Science* 271:1705-1707.

Chen, J., M.A. Reed, A.M. Rawlett, and J.M. Tour. 1999. Large on-off ratios and negative differential resistance in a molecular electronic device. *Science* 286:1550-1552.

Chou, S., and P.R. Krauss. 1996. Quantum magnetic disk. *J. Magn. Magn. Mater.* 155:151.

Collier, C.P., E.W. Wong, M. Belohradský, F.M. Raymo, J.F. Stoddart, P.J. Kuekes, R.S. Williams, and J.R. Heath. 1999. Electronically configurable molecular-based logic gates. *Science* 285:391-394.

Collins, P.G., A. Zettl, H. Bando, A. Thess, and R.E. Smalley. Nanotube nanodevice. *Science* 278:100.

Cuberes, M.T., et al. 1996. Room temperature repositioning of individual C60 molecules at Cu steps: Operation of a molecular counting device. *Appl. Phys. Lett.* 69:3016.

Credi, A., V. Balzani, S.J. Langford, and J.F. Stoddart. 1997. Logic operations at the molecular level. An XOR gate based on a molecular machine. *J. Am. Chem. Soc.* 119:2679.

Disktrend. 1998. http://www.disktrend.com.

DiVincenzo, D. 1995. Quantum computation. *Science* 270:255.

Dong, L.F., et al. 1997. Gas sensing properties of nano-ZnO prepared by arc plasma method. *Nanostruct. Mater* 8:815.

Duncan, R. 1997. Polymer therapeutics for tumor specific delivery. *Chemistry and Industry* 7:262-264.

Ellenbogen, J.C. and J.C. Love. 1999. *Architectures for molecular electronic computers: 1. Logic structures and an adder built from molecular electronic diodes*. McLean, VA: The MITRE Corporation, Report MP 98W0000183. July. See also references cited in the extensive bibliography in this work. (Available on the Internet at: http://www.mitre.org/technology/nanotech/.)

Gershenfeld, N., and I.L. Chuang. 1997. Bulk spin resonance quantum computation. *Science* 275:350.

Gimzewski, J.K., et al. 1998. Rotation of a single molecule within a supramolecular bearing. *Science* 281:531.

Grochowski, E. 1998. Emerging trends in data storage on magnetic hard disk drives. *Datatech*, 11.

Guo, L.J., et al. 1997. A single electron transistor memory operating at room temperature. *Science* 275:649.

Guo, T. et. al. 1995. Catalytic growth of single walled nanotubes by laser vaporization. *Chem. Phys. Lett.* 243:49.

Gurney, B., and E. Grochowski. 1998. Spin-valve sensors take up where MR heads leave off. *Data Storage,* September, 59.

Hasslacher, B., and M.W. Tilden. 1995. Living machines. In *Robotics and autonomous systems: The biology and technology of intelligent autonomous agents.* Ed. L. Steels. Elsevier (LAUR-94-2636).

Heath, J.R., P.J. Kuekes, G.S. Snider, and R.S. Williams. 1998. A defect-tolerant computer architecture: Opportunities for nanotechnology. *Science* 280:1716-1721.

Iijima, S. 1991. Helical microtubules of graphitic carbon. *Nature* 354:56.

IPGA. N.d. http://ipga.phys.ucl.ac.uk/research/arrays/rtt-paper.html#21.

IPT. N.d. http://www.ipt.arc.nasa.gov.

Jung, T.A., et al. 1996. Controlled room-temperature positioning of individual molecules: Molecular flexure and motion. *Science* 271:181.

Kergueris, C., et al. 1999. Electron transport through a metal-molecule-metal junction. *Phys. Rev. B* 59:12505.

Kim. J., et al. 1999. A single-photon turnstile device. *Nature* 397:500.

Lee, S.C. 1998. The nanobiological strategy for construction of nanodevices. In Lee, S.C. and L. Savage (eds.), *Biological molecules in nanotechnology: the convergence of biotechnology, polymer chemistry and materials science*. Southborough, MA: IBC Press.

Lent, C., and D. Tougaw. 1997. A device architecture for computing with quantum dots. In *Proc. IEEE* 85:541.

Leobandung, E., et al. 1995. Observation of quantum effects and Coulomb blockade in silicon quantum dot transistors at temperatures over 100K. *Appl. Phys. Lett.* 67:938.

Lutwyche, M., et al. 1998 . Microfabrication and parallel operation of 5x5 AFM cantilever arrays for data storage and imaging. In *Proc. MEMS98, IEEE 11th Annual International Workshop on MEMS*, 8.

Lyo, I-.W., and Avouris, P. 1991. Field-induced nanometer to atomic-scale manipulation of silicon surfaces with the STM. *Science* 253:173.

Mao, C., et al. 1999. A nanomechanical device based on the B-Z transition of DNA. *Nature* 397:144.

Martel, R., T. Schmidt, H.R. Shea, T. Hertel and P. Avouris. 1998. Single- and multi-wall carbon nanotube field effect transistors. *Appl. Phys. Lett.* 73(17):2447.

Matsumoto, K., et al. 1996. Room temperature operation of a single electron transistor made by the STM nanooxidation process for the TiOx/Ti system. *Appl. Phys. Lett.* 68:34.

Minne, S.C., et al. 1996. Independent parallel lithography using the AFM. *J. Vac. Sci. Technol B* 14:2456.

Moffat, C. 1999. University College, London (personal communication).

Mucic, R.C., et al. 1998. DNA-directed synthesis of binary nanoparticle network materials. *J. Amer. Chem. Soc.* 120:12674.

Murray, C.B., et al. 1995. Self organization of CdSe nanoparticles into three dimensional quantum dot superlattices. *Science* 270:1335.

Noji, H., et al. 1997. Direct observation of the rotation of F1-ATPase. *Nature* 386:299.

Orlov, A.O., et al. 1997. Realization of a functional cell for quantum dot cellular automata. *Science* 277:928.

Petty, M.C., M.R. Bryce and D. Bloor, eds. 1995. *An introduction to molecular electronics*. New York: Oxford University Press.

Pomrenke, G.S. N.d. http://web-ext2.darpa.mil/mto/ULTRA.

Prinz, G.A. 1998. Device physics—magnetoelectronics. *Science* 282:1660.

Reed, M.A., et al. 1997. Conductance of a molecular junction. *Science* 278:252.

Requicha, A. 1999. Nanorobotics on surfaces. In *IWGN Workshop Proceedings*, January 27-29, 1999. (personal communication).

Resch, R., et al. 1998. Manipulation of nanoparticles using dynamic force microscopy: Simulation and experiment. *Appl. Phys. A* 67:265.

Seabaugh, A. 1998. http://www.darpa.mil/mto/ultra/98Overview/Raytheon-29.html.

Shor, P. 1994. Algorithms for quantum computation: Discrete logarithms and factoring. In *Proc. 35th Annu. Symp. Foundations of Computer Science*, 124.

Sievenpiper, D.F., et al. 1998. 3D metallo-dielectric photonic crystals with strong capacitive coupling between metallic islands. *Phys. Rev. Lett.* 80:2829.

Spudich, J.A. 1994. How molecular motors work. *Nature* 372:515.

Stroscio, J.A., and Eigler, D. 1991. Atomic and molecular manipulation with the STM. *Science* 254: 1319.

Tans, S.J., et al. 1997. Individual single-wall carbon nanotubes as quantum wires. *Nature* 386:474.

Tilden, M.W. 1998. Autonomous biomorphic robots as platforms for sensors. Report Number 'LA-UR--96-3222. U.S. Information Bridge, DOE Office of Scientific and Technical Information (OSTI). http://www.doe.gov/EnergyFiles/, Jan 21.

_____. 1999. Analysis of living biomech machines using minimal chaotic control. In *AROB 99 proceedings, The Fourth International Symposium on Artificial Life and Robotics*. University of Beppu, Japan, Jan 19: 99-101.

Zhou, C., M.R. Deshpande, M.A. Reed, L. Jones II, and J.M. Tour. 1997. Nanoscale metal/self-assembled monolayer/metal heterostructures. *Appl. Phys. Lett.* 71:611.

Chapter 7

APPLICATIONS: CONSOLIDATED NANOSTRUCTURES

Contact persons: R.W. Siegel, Rensselaer Polytechnic Institute;
B. Kear, Rutgers University

7.1 VISION

Nanostructure science and technology (nanotechnology) is fundamentally changing the way that materials and the structures made from them will be manufactured in the future. Our increasing ability to synthesize and assemble nanoscale building blocks with precisely controlled sizes and chemistries into consolidated nanostructures and nanocomposites with unique properties and functionalities likely will lead to revolutionary changes in industry.

7.2 CURRENT SCIENTIFIC AND TECHNOLOGICAL ADVANCEMENTS

A number of major scientific and technological advances in the area of consolidated nanostructures have occurred in the past decade, and many more are expected. Such advances, some of which are listed below, have already led to commercial scale-up of some nanostructured materials and also to products incorporating them:

- Production of layered nanostructures with control of thickness at the atomic level and the subsequently developed ability to engineer the resisto-magnetic-field response by varying nanoscale architecture to make useful devices for magnetic recording
- Development of processes to net-shape-form nanophase ceramics and ceramic-based composites into finished parts while maintaining ultrafine grain size and nanoscale properties when desired

- Discovery and development of unique nanostructured hard and soft magnetic materials for a variety of applications, including information technology hardware
- Development of nanoscale cemented-carbide, hard materials for improved cutting-tool performance with superior wear resistance and fracture toughness
- Development of direct methods for fabricating nanostructured coatings yielding exceptional electrical, chemical, thermal, mechanical, and environmental protection of the coated parts
- Creation of a wide range of nanocomposites, such as nanoparticle- or nanotube-filled polymers, with enhanced or fundamentally new and controllable engineering performance, including significantly increased strength and reduced flammability
- Development of biological templating for the directed growth and patterning of nanostructures for biomedical and electronic applications
- Engineering of scaled-up and economical industrial processes for production of nanopowders and nanostructured bulk materials in the multi-tonnage range

7.3 GOALS FOR THE NEXT 5-10 YEARS: BARRIERS AND SOLUTIONS

In order to move nanostructure science and technology forward effectively during the next five to ten years, there are a number of significant issues that must be considered.

Significant Issues

- Nanostructured systems will be manufactured primarily from the bottom up using nanoscale building blocks and not from the top down as in conventional manufacturing methods.
- Up-front costs may be higher, but net commercial savings will result from more efficient design and processing, use of only necessary material, and decreased and less deleterious effluent.
- New scale-up methods will be required, along with novel scenarios for processing and handling of materials and parts.

- Statistically driven process controls with real time diagnostics and precise reliability standards will also be needed.
- Realistic multiscale theoretical modeling of nanostructuring and the properties and functionalities of the resulting nanostructures and systems will be important to their ultimate success.
- Understanding of the structure and properties of surfaces and interfaces and how to control these in a variety of nanostructure assembly strategies will need to be significantly increased.

Many new applications in the area of consolidated nanostructures will become available when these issues are faced and the appropriate barriers surmounted. Some examples follow:

- Ultrahigh-strength, tough structural materials
- Novel soft and hard ferromagnets
- Ductile and strong cements
- High-brightness displays based on nanotubes
- Bio-inspired medical prostheses
- Self-assembled arrays of biomolecular single-electron devices
- Drugs from consolidated nanoparticles

Several additional opportunities can be suggested that will lead to a wide range of useful developments:

- High-pressure sintering of nanophase oxide ceramics should be extended to difficult-to-sinter non-oxide ceramics for a variety of industrial applications, including high-efficiency automotive parts.
- Magnetic nanocomposites with 5-10-fold increases in magnetocaloric effects should be used to develop magnetic refrigerators that operate at room temperature.
- A unique new class of nanostructured permanent magnets with higher saturation magnetization values than are presently available should be developed for many information technology applications.

- Textiles and plastics filled with dispersed nanoparticles should be created that will enhance the materials' mechanical performances while reducing their flammability.
- Lithium-ion batteries composed of nanodispersed ceramics in polymer matrices should be developed for the next generation of lightweight rechargeable batteries for cellular telephones, laptop computers, and transportable CD players and radios.

7.4 SCIENTIFIC AND TECHNOLOGICAL INFRASTRUCTURE

In order for nanostructure science and technology, including the area of consolidated nanostructures, to reach fruition in the coming years, a number of infrastructure issues need to be addressed:

- Research funding in nanotechnology should focus on individuals or groups of researchers in universities with interdisciplinary interests, alone or coupled with industry in research/training partnerships, with the national laboratories providing support through unique major facilities and capabilities.
- The research funding process needs to be streamlined by reducing agency timelines, minimizing the proposal writing and reviewing logjam, and increasing significant inter- and intra-agency funding efforts, for example with shorter proposals and greater use of panel reviews.
- It is imperative to create a new breed of researchers who can think "outside the box" of traditional disciplines.
- Educating this new breed of researchers, who will either work across disciplines or know how to work with others across disciplinary lines in the interfaces between disciplines, is vital to the future of nanotechnology.
- Significant opportunities need to be created for academic training, to include industrial internships.

7.5 R&D INVESTMENT AND IMPLEMENTATION STRATEGIES

Besides the single investigator and group research programs, a national nanotechnology network should be established between existing and new nodal points at university centers, while at the same time extending the links to industry and the national laboratories. Research and training in nanostructure science and technology should be given high priority.

University nanostructure science and technology centers should be funded at a level in the range of $1-5 million per year, with opportunities for renewal after 5 years. Matching funds from industrial partners must be a strong consideration. Additional funds should be earmarked annually in the Federal budget for these nanotechnology centers, without decreasing the normal program funding in this area.

The National Science Foundation's Grant Opportunities for Academic Liaison with Industry (GOALI) program provides an excellent model for university-industry collaborations in fundamental research. The Small Business Innovative Research (SBIR) and Small Technology Transfer Research (STTR) programs supported by Federal agencies remain attractive means for promoting technology transfer and implementation to put university research into industrial practice. The Department of Commerce's Advanced Technology Program (ATP) is another example of targeting research on industrial needs.

7.6 PRIORITIES AND CONCLUSIONS

The area of consolidated nanostructures has seen a great deal of progress in the past decade, with several commercially scaled-up processes now a reality, new jobs created, and new applications already in the marketplace. These successes have been based on the scientific and technological advances that have resulted largely from fundamental research, which then moved toward directed research and development, and from there to manufacturing. The benefits to be gained in the future are enormous compared to those that have been reaped to date. We must focus significant new resources to nurture the fundamental research that will provide the seedbed for future new ideas in consolidated nanostructures; at the same time we must develop new opportunities for training the scientists and engineers needed in this vibrant new field and providing a well lubricated path for technology transfer to industry and the marketplace.

7.7 EXAMPLES OF CURRENT ACHIEVEMENTS AND PARADIGM SHIFTS

Five detailed examples are presented here to elucidate some of the successes and considerable opportunities in the area of consolidated nanostructures. These include various nanostructured layers used in the creation of novel giant magnetoresisitance devices for information technology hardware; nanoscale composites of ceramics and metals (cermets) used for hard coatings or wear parts in drilling operations; a novel nanoscale processing route for making a wide variety of ceramic parts; and well-dispersed nanoparticles used for flame retardation in plastics for transportation systems. Some of the examples are already commercial, while others are moving rapidly in that direction.

7.7.1 Nanostructures Used for Giant Magnetoresistance (GMR) Devices

Contact person: R. Shull, National Institute of Standards and Technology

Giant magnetoresistance (GMR) is a nanoscale phenomenon first discovered in 1988 by A. Fert and coworkers in alternating nanoscale multilayers of Fe and Cr (Baibich et al. 1988). A number of material pairs consisting of a strong ferromagnet (e.g., Fe, Co, NiFe) and a weaker magnetic or nonmagnetic buffer (e.g., Cr, Cu, Ag) have been found to exhibit a similar response magnetoresistance effect in multilayers. The effect results from variations in electron scattering from the interlayer interfaces as an external magnetic field is applied parallel to the multilayers. The spins in alternate magnetic layers in the absence of an external magnetic field are oppositely aligned through anti-ferromagnetic coupling and yield maximum scattering, while in a sufficiently strong positively or negatively oriented external magnetic field, these spins align with the field, and hence, with one another, decreasing the scattering at the interfaces. However, it has been subsequently observed that such sets of alternating, or one-dimensionally modulated, nanolayers are not the only architecture that exhibits GMR. Random distributions of spherical magnetic nanoparticles in a nonmagnetic matrix give similar results, although the rather large magnetic fields necessary to obtain a minimum resistance in all of these consolidated nanostructures are too large for most practical applications. The discovery that changing the modulation dimensionality of the GMR nanoarchitecture to three dimensions could lead to great reductions in the required magnetic fields has opened the way to the creation of a wide variety of useful GMR devices, some of which are shown schematically in Figure 7.1 (Barthélémy et al. 1994). Several such devices are now used in commercial information technology hardware, especially in magnetic reading devices.

7.7.2 Nanostructured Hard Materials

Contact person: B.H. Kear, Rutgers University

Nanostructured hard materials are beginning to have commercial impact. It has become possible during the past decade to create cemented carbide nanocomposites, such as WC/Co and TiC/Fe, that have considerably enhanced hardness, fracture toughness, and wear resistance compared to their conventional grain size counterparts that are widely used in the manufacture of machine tools, drill bits, and wear parts. The exceptional properties of nanostructured materials are realized when the constituent WC and Co phases, for example, are interconnected in three dimensions, forming a so-called bicontinuous nanostructure. Typically, WC/Co products are produced by mechanically mixing powders of the constituent phases, followed by cold pressing and sintering. The difficulty of uniformly mixing ultrafine WC and Co powders by mechanical means has heretofore limited the scale of the WC grain size attainable in the final sintered product to about 300 nm. Over the past several years, new chemical methods have been developed for producing pre-mixed powders at the nanoscale level.

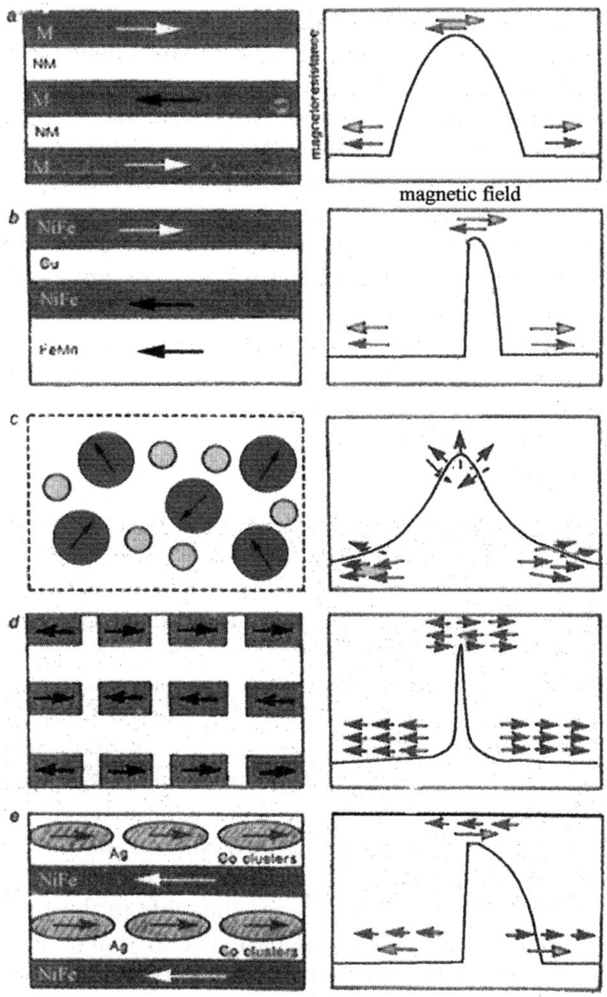

Figure 7.1. Various GMR nanostructures (left) and their magnetoresistance behavior (right—note that all the horizontal scales are different): (a) anti-ferromagnetically coupled multilayer; (b) spin-valve structure; (c) granular alloy; (d) multilayer with discontinuous magnetic layers; and (e) hybrid nanostructure including clusters and layers (reprinted by permission from *Physics World*, Barthélémy et al. 1994).

As an example, in one process that has been scaled up to produce tonnage quantities, spray drying is used to produce a homogeneous precursor powder of mixed tungsten and cobalt salts, followed by fluid-bed thermochemical conversion (pyrolysis, reduction, and carburization) to transform the precursor powder into the desired nanophase WC/Co product powder. Typically, the WC particle size in the agglomerated powder product is about 30 to 40 nm. Using liquid phase sintering and the addition of small amounts of a grain growth inhibitor such as VC, fully sintered nanocomposite products can be produced without inducing significant coarsening of the WC grains. Several hard metal companies are now using these powders in the manufacture of high performance parts, such as microtwist drills used for drilling holes in printed circuit boards (shown below in Figure 7.2); this is an industrial process in which drill tip wear has been a major problem.

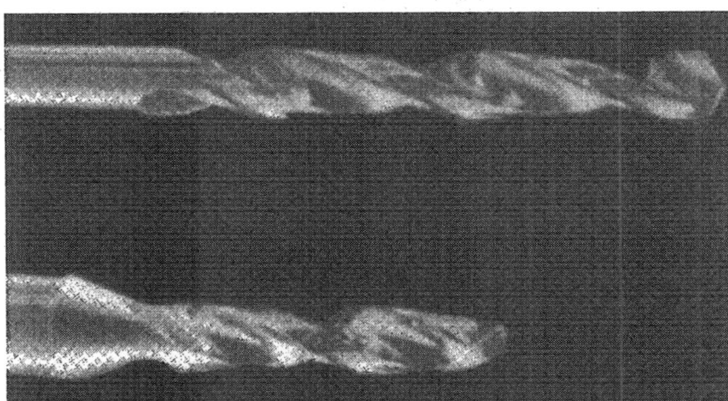

Figure 7.2. Uses for consolidated nanostructured hard materials: nanocomposite microtwist drill bit (top) compared to conventional product after wear for the same use time (courtesy Nanodyne Inc.).

In addition, with the recent availability of industrial-scale quantities of nanopowders of cermets and also ceramics, the use of nanostructured materials as feedstocks for thermal spraying of nanostructured coatings onto conventional parts is being explored through an initiative of the Office of Naval Research (ONR) to implement the new technology into shipboard systems. Preliminary work has demonstrated that these coatings exhibit superior abrasion and corrosion wear resistance in diverse applications, which could lead to early introduction of such coatings into naval construction.

7.7.3 Ceramic Nanoparticles

Contact person: R.W. Siegel, Rensselaer Polytechnic Institute

A new processing route toward the economic manufacture of ceramic parts has been enabled by ceramic nanoparticles in the past few years. The ability to net shape form ceramics into final parts (as shown below in Figure 7.3) has become a reality in recent years, owing to advancements made in the scaled-up production and processing of ceramic nanoparticles. With the availability now of tonnage quantities of nanophase ceramic powders with their unique rheological and mechanical properties, it is possible to directly form ceramic parts in a sinter-forging mold under sufficient pressure and temperature to yield final parts with all of the definition and precision of the original mold. Nanophase ceramics such as titania and alumina, made from the consolidation of ceramic nanoparticles, have been shown in the laboratory to be readily formable into small samples, and studies of their mechanical behavior indicate that a significant degree of ductile behavior in compression is exhibited in these ultrafine grain size materials.

Figure 7.3. Net shape forming via consolidated nanoparticles (courtesy of Nanophase Technologies Corporation).

An understanding of this behavior has been developed in terms of earlier models of grain boundary sliding, in which grains (formerly nanoparticles before consolidation) can slide over one another under the influence of an applied forming stress without breaking the bonds across the boundaries between grains as a result of diffusional (atom transport) healing events taking place at any incipient cracks. This diffusional accommodation in the grain boundary regions over the very short distances involved in nanoscale materials is the key to enabling the net shape forming of nanophase ceramics and nanocomposites based on them. While still in the early stages of commercial activity, such net shape forming technology should have significant future impact.

7.7.4 Fire Retardation in Plastics

Contact person: J.W. Gilman, National Institute of Standards and Technology

Plastics have rapidly infiltrated all aspects of society and have enabled many technologies that are commonplace today, such as aircraft, automobiles, and many consumer goods. However, beyond the many benefits of plastics, their inherent flammability is a continual concern in many applications. In addition to the hazard from fire, the concomitant smoke and toxic combustion products are a serious problem. The flammability of plastics can be considerably reduced by the addition of well-dispersed inorganic nanoparticles to form consolidated nanocomposites. For example, the heat release rates of thermoplastic and thermoset polymer-based materials are reduced by 40-60% in delaminated or intercalated nanocomposites that contain only 2-6 wt% silicate clays.

An example is shown in Figure 7.4. The nanocomposite structure of the char appears to enhance the performance of the char layer, which apparently acts as an insulator and a mass transport barrier, slowing the escape of the volatile products generated as the polymer decomposes. This marked reduction in flammability results in self-extinguishing characteristics and is achieved without the commonly observed decrease in mechanical properties and increase in evolution of carbon monoxide, soot, and toxic compounds associated with conventional flame-retardant additives: physical properties are not degraded, and processability is maintained. Beyond their flammability resistance, such nanocomposites have also demonstrated ablation resistance comparable to current state-of-the-art solid rocket motor ablatives. Ablative materials are critical for insulation applications in space and launch systems in order to protect aerodynamic surfaces, propulsion structures, payloads, and ground equipment from the severe effects of very high temperatures (>2000°C), incident heating rates, and chemically oxidizing atmospheres with gas and particulate velocities that may range from Mach 0.01 to 10+. As little as 1.6 vol% of well-dispersed nanoscale layered silicate in nylon 6 is needed to form a relatively tough, inorganic char during ablation. This protective char results in at least an order-of-magnitude decrease in the mass loss (erosion) rate relative to neat nylon 6.

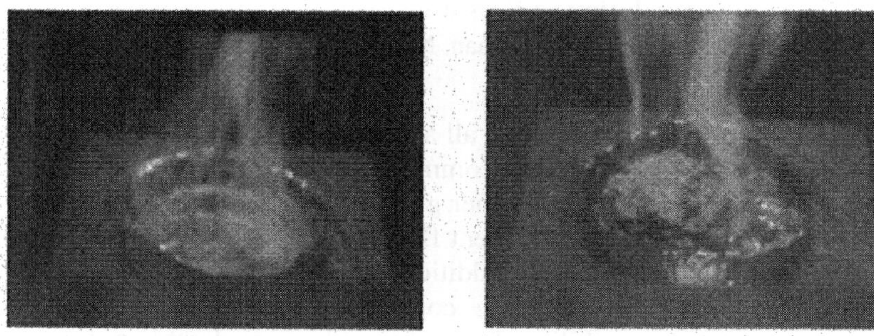

Figure 7.4. Flammability and thermal stability studies of polymer layered silicate (clay) nanocomposites (Gilman 1999; reproduced by permission).

7.7.5 Manufacturing of Nanostructured Coatings

Contact person: E. Lavernia, University of California, Irvine (UCI)

Thermal spraying of nanostructured coatings represents a potentially revolutionary approach to capitalizing on the unusual mechanical and physical attributes of nanostructured materials (e.g., hardness, toughness, and resistance to corrosion). When grain size is in the nanoscale range, the number of atoms at the grain boundaries becomes comparable to the number of atoms inside the grains. Because specific grain boundary area increases, the amount of impurities per unit grain boundary is lower than for larger-grained materials with the same bulk impurity concentration. This purification of the grain boundaries has been associated with more uniform corrosion morphology and higher intergranular corrosion resistance than larger-grained materials.

Recent work (Lau, Jiang, et al. 1998; Lau, Strock, et al. 1998) has demonstrated that it is possible to achieve dramatic enhancement in the physical behavior of nanostructured coatings, that is, coatings having grain sizes smaller than 100 nanometers (e.g., see Figure 7.5). Available results suggest that the nanometric grains are not only inherently thermally stable (Huang et al. 1996), but they also effectively block dislocation movement, giving rise to ultrahigh hardness, and in some cases, ultrahigh toughness values. Additional benefits of the nanometric grain sizes of the coatings include a reduction in the residual stress state of the coatings, which effectively allows, for the first time, the generation of thicknesses heretofore unobtainable—in some cases, four times thicker than those achievable with conventional materials. For example, various nanostructured coatings (Ni, Ni-based superalloy and stainless steel; Cr_3C_2/NiCr; and WC/Co) have been successfully thermal sprayed in UCI laboratories using high velocity oxygen fuel (HVOF) spraying. Comparison of the nanocrystalline coating characteristics to those of conventional coatings showed larger micro-hardness values for the as-sprayed nanocrystalline coatings than for the conventional coatings, an increase of 16 to 63%, depending on the gas composition and the milling method.

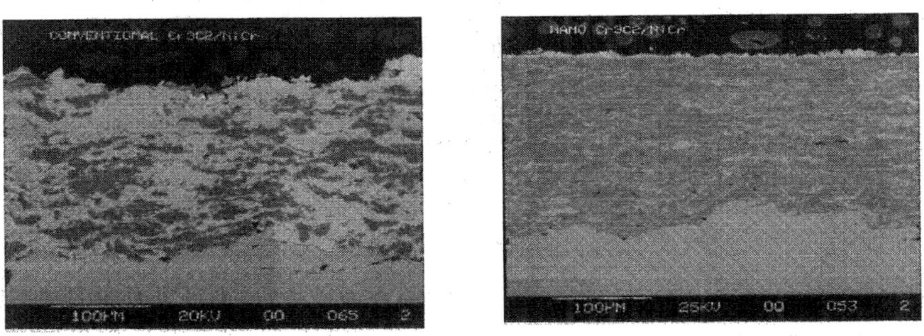

(a) Conventional Cr_3C_2/NiCr coating (b) Nanostructured Cr_3C_2/NiCr coating

Figure 7.5. SEM morphology of conventional and nanostructured coatings.

Potential applications span the entire spectrum of technology, from thermal barrier coatings for turbine blades to wear-resistant rotating parts. The potential economic impact is several billion dollars per year, and development of this technology will likely involve the aerospace industry (e.g., Boeing Corporation), the jet engine industry (e.g., General Electric), and the automotive industry (e.g., Ford) (Cheung et al. 1996, 479).

7.8 REFERENCES

Baibich, M.N., J.M. Broto, A. Fert, F. Nguyen Van Dau, F. Petroff, P. Etienne, G Creuzet, A. Friederich, and J. Chazelas. 1988. Giant magnetoresistance of (001)Fe/(001)Cr magnetic superlattices. *Phys. Rev. Lett.* 61:2472.

Barthélémy, A., A. Fert, R. Morel, and L. Steren. 1994. Giant steps with tiny magnets. *Physics World*. November. Vol. 7, No. 11: 34-38.

Cheung, C., D. Wood, and U. Erb. *1996.* In *Processing and properties of nanocrystalline materials*, ed. C. Suryanarayana, J. Singh, and F.H. Froes. Warrendale, Pennsylvania: The Minerals, Metals and Materials Society.

Eastman, J. 1999. Nanostructured materials at Argonne National Laboratory. In *IWGN Workshop Proceedings*, January 27-29, 1999. (personal communication).

Gilman, J.W. 1999. Flammability and thermal stability studies of polymer layered-silicate (clay) nanocomposites. *Applied Clay Science* 15:31-49.

Helms, J.H. 1999. Nanotechnology in the automotive industry. In *IWGN Workshop Proceedings*, January 27-29, 1999. (personal communication).

Huang, B., R.J. Perez, H. Hu, and E.J. Lavernia. 1996. Grain growth of nanocrystalline Fe-Al alloys produced by cryomilling in liquid argon and nitrogen. *Materials Science and Engineering A* 255: 124-132.

Lau, M.L., H.G. Jiang, W. Nuchter, and E.J. Lavernia. 1998. Thermal spraying of nanocrystalline Ni coatings. *Solidi Status Physica,* 257-268.

Lau, M.L., E. Strock, A. Fabel, C.J. Lavernia and E.J. Lavernia. 1998. Synthesis and characterization of nanocrystalline Co-Cr coatings by plasma spraying. *NanoStructured Materials* 10(5):723-730.

Chapter 8

APPLICATIONS: BIOTECHNOLOGY, MEDICINE, AND HEALTHCARE

Contact persons: H. Craighead, Cornell University; K. Leong, Johns Hopkins University

8.1 VISION

Nanotechnology is beginning to allow scientists, engineers, and physicians to work at the cellular and molecular levels to produce major benefits to life sciences and healthcare. In the next century, the emerging field of nanotechnology will lead to new biotechnology-based industries and novel approaches in medicine.

8.2 CURRENT SCIENTIFIC AND TECHNOLOGICAL ADVANCES

Major advances in the last several years in scanning probe and scanning optical analytical methods permit viewing the vital chemical processes and microscopic structures in biological systems with unprecedented resolution. These new analytical probes reveal a detailed picture of the microscopic structure of living cells and a view of chemical processes at the molecular scale. The atomic force microscope, for example, can locate and measure the extraordinarily small forces associated with receptor-ligand binding on cell surfaces. Microscopic electrical probes can detect a living cell's exchange of ions with its environment or the propagation of electrical signals in nerves. New high-resolution optical instruments, combined with chemically selective light-emitting fluorescent probes, can follow in detail the chemical processes on the surface of and inside a living cell. This analytical capability allows observation of the biochemical processes and interactions of cells in living systems.

Cells contain exquisite naturally occurring "molecular motors." One of many examples of these naturally occurring nanomachines is F1-ATPase, which is part of the large, membrane-embedded complex that synthesizes ATP within mitochondria (Figure 8.1). This structure, only about 10 nm in size, is a robust, fully functional rotating motor that is powered by natural biochemical processes. In 1998 the Amersham Pharmacia Biotech and Science Prize was awarded to Hiroyuki Noji, a young Japanese scientist who demonstrated the function of this molecular motor by attaching a long actin filament to the rotating part of the motor and observing the rotation in an optical microscope. The detailed understanding of the structure and function of this motor protein and other macromolecular assemblies essential for life is an area of growing scientific importance.

During the last few years, scientists have developed the technology for rapidly mapping the genetic information in DNA and RNA molecules, including detection of mutations and measurement of expression levels. This technology uses DNA microchip arrays that adapt some of the lithographic patterning technologies of the integrated circuit industry.

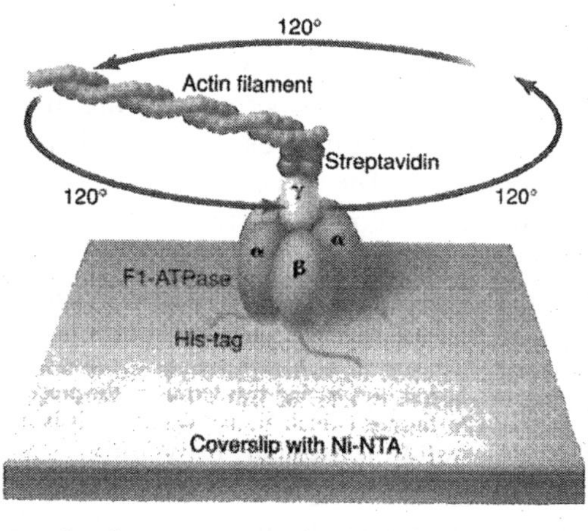

Figure 8.1. The molecular motor protein F1-ATPase. Illustrated here is an experiment reported in *Science*, in which an actin filament is attached to a motor protein to provide load to and allow visualization of the motor rotation (reprinted with permission from Noji 1998, ©1998 American Association for the Advancement of Science).

This is now a commercial technology and is finding its way into biotechnology research and industrial utilization. Work on new types of chemical arrays should expand this approach of parallel biological information processing to analysis of proteins and other biomolecules. Miniaturization of allied analytical processes such as electrophoresis will lead to increases in throughput and reduced cost for other important methods of analysis such as DNA sequencing and fingerprinting. For example, new research (Turner et al. 1998) is aimed at replacing the tedious, slow, and expensive process of DNA sequencing in slab gels with miniaturized integrated microfabricated analytical systems (Figure 8.2).

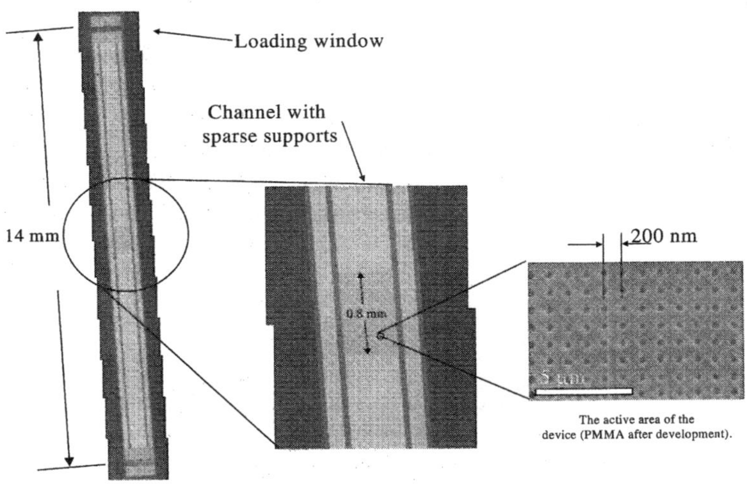

Figure 8.2. Photomosaic of a DNA separation chip. The image is pieced together from twelve optical micrographs. The inset shows a small region 0.8 mm long containing dense pillars that act as a molecular sieve to separate DNA molecules according to size. Conventional gel electrophoresis works essentially the same way, and for this reason these nanofabricated structures are called "artificial gels." This technology, while far from commercialization, has the potential to revolutionize DNA separation techniques by providing an inexpensive, durable, and reproducible medium for DNA electrophoresis (courtesy S.W. Turner, Cornell Univ.).

Using biological systems as a model, scientists are attempting to build ever more complex systems that are capable of self-assembly. As the sizes of components become smaller and manipulation of these components becomes impracticably slow, the need for self-assembling systems is rising. Complex biological systems provide models from which to design components that can come together in only one way to form the desired three-dimensional nanoarchitectural system. Similarly, scientists are using strategies learned from biological systems to design new materials. Spider silk is one of the strongest materials known. Its molecular structure is being used to design better composite polymer systems of increasing strength and utility.

Nanoparticles considerably smaller than one micron in diameter have been used in revolutionary ways to deliver drugs and genes into cells. The particles can be combined with chemical compounds that are ordinarily insoluble and difficult for cells to internalize. The derivatized particles can then be introduced into the bloodstream with little possibility of clogging the capillaries and other small blood vessels, as in the case of insoluble powders. The efficacy and speed of drug action in the human body can thereby be dramatically enhanced. In similar ways, nanoparticles carrying DNA fragments can be used to incorporate specific genes into target cells (Figure 8.3).

Figure 8.3. Pictured here is the "Gene Gun," a system that uses nanoparticles to deliver genetic material to transfect plant and animal cells. In this system, submicron gold particles coated with DNA are accelerated with a supersonic expansion of helium gas. The particles leave the front of the device at high velocity and penetrate the cell membrane and nuclear membrane, thus delivering the genetic material to the nucleus (courtesy Bio-Rad Laboratories).

The ability of DNA to undergo highly controlled and hierarchical assembly makes it ideal for applications in nanobiotechnology. For example, DNA has been used to design lattices that readily assemble themselves into predictable, two-dimensional patterns. These arrays are composed of rigid DNA tiles, about 60 nm^2, formed by antiparallel strands of DNA linked together by a double-crossover motif analogous to the crossovers that occur in meiosis. The precise pattern and periodicity of the tiles can be modified by altering DNA sequence, allowing the formation of specific lattices with programmable structures and features at a nanometer scale. This approach has the potential to lead to the use of designed DNA crystals as scaffolds for the crystallization of macromolecules, as materials for use as catalysts, as molecular sieves, or as scaffolds for the assembly of molecular electronic components or biochips in DNA-based computers. Similarly, biological-molecule-based scaffolding could take advantage of the unique structural characteristics of RNA molecules, of polypeptide chains, or of the highly specific interactions that occur between DNA and proteins or between RNA and proteins.

Devices that are currently in use to control the interactions of DNA on surfaces can have broader applications for controlling nanoassembly. These devices use electric fields to control the movement of particles toward or away from microscopic sites on the device surface. Charged biological molecules (DNA, RNA, protein) and analytes, cells, and other nanoscale or microscale charged particles can be precisely organized.

8.3 GOALS FOR THE NEXT 5-10 YEARS: BARRIERS AND SOLUTIONS

The advances noted above and others involving nanofabrication and nanosynthesis are enabling significant new opportunities for scientific research and commercial applications.

The integration and miniaturization of fluid control, or fluidics, with photonics and electronics is a trend that will lead to a paradigm change in chemical synthesis and analysis. Industries that have not previously been considered high-tech will be transformed by nanofabrication technology in the twenty-first century.

Given the inherent nanoscale of receptors, pores, and other functional components of living cells, the detailed monitoring and analysis of these components will be made possible by the development of a new class of nanoscale probes. Nanotechnology will improve the sensitivity and integration of analytical methods to yield a more coherent evaluation of life processes. The ability to manipulate cells and integrate them with complex inorganic devices and probes will permit scientists to perform a new class of experiments and ask new questions about basic cell functions. For example, integrated cellular systems grown in culture could replace and thus spare animals used for testing drugs and hazardous materials.

Nanoscale sensors. Integrated nanoscale sensors could monitor the condition of a living organism, the environment, or components of the nutrient supply, sampling a range of conditions with a high degree of sensitivity. With arrays of ultraminiaturized sensors that sample a range of chemicals or conditions, the confidence level and specificity of detection would be much greater than is now possible with separate macroscopic sensors. As has been seen with electronic integrated circuits, as the level of device integration increases and the volume of production grows, the costs of highly complex units decreases. One can project that in the next century highly sophisticated, small, and inexpensive sensors employing nanotechnology will be available and used routinely in many parts of our lives.

Nanomachines. To date, development of miniaturized devices is based mostly on nonbiological principles. An example of an autonomous miniaturized controlled-released implantable device (a solid-state silicon microchip) for drug delivery applications is illustrated in Figure 8.4. The microchip can release a single or multiple chemical substance(s) on demand. In addition to drug delivery, this technology may also find use in such areas as diagnostics, analytical chemistry, and others.

As integrated nanofabricated systems decrease in size, the ability to retain desired functions will become more difficult. As has been noted, nature has solved many of these same engineering problems and has produced functional molecular motors and many other subcellular functional machines. Further research should allow scientists to integrate these natural systems with inorganic devices and create hybrid systems and a new class of nanomechanical devices.

Nanomachines powered by chemically fueled molecular motors could be coupled to devices with integrated valves, pumps, and sensors that can react to changes in the body and the environment. One can imagine, for instance, miniaturized, self-powered machines that sense and identify oil or chemical pollutants in soils and map their distribution and concentration, or medical implants that sense and dispense drugs or hormones in response to body changes.

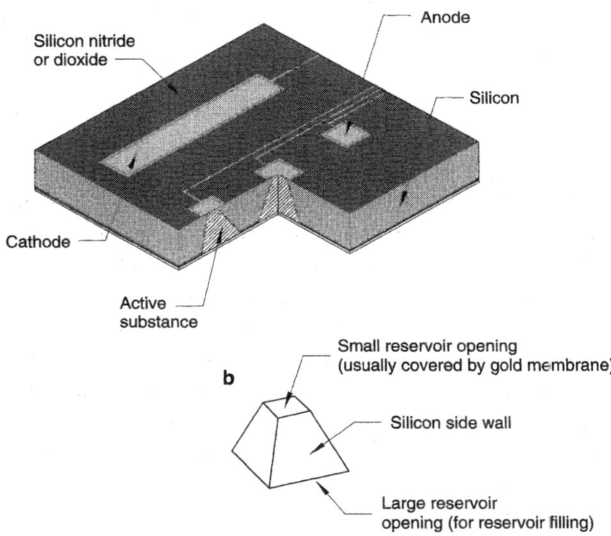

Figure 8.4. Prototype of a microchip device for drug delivery (reprinted by permission from *Nature*, Santini et al. 1999, ©1999, Macmillan Magazines Ltd.).

Nanoparticles. Current bioengineered, non-viral gene vectors that are used to introduce new genes into cells are far from perfect. Ideally, DNA nanoparticles with controlled composition, size, polydispersity, shape, morphology, stability, encapsulation capability, and targetability will result in new technologies with improved *in vivo* transfection efficiency. Such nanotechnology will likely have a significant impact on realizing the potential of genetic engineering techniques in agriculture, manufacturing, and environmental applications, as well as in medicine.

Drug development. Technology is dramatically accelerating the discovery of new drug compounds. Continuing advances in nanotechnology will lead to innovative synthetic routes, new processing strategies, and more economical manufacturing. The same or similar processes that have led to the phenomenal increases in computational speed of microprocessors and the increasing density of computer memory will similarly revolutionize the speed with which new compounds are screened for therapeutic potential as new drugs. The pharmaceutical industry projects nearly a tenfold increase in the number of drug compounds that will be evaluated in 2000 compared to 1998, with only a modest miniaturization of technology. If the trend is similar to that of microelectronics, the rate could grow exponentially. Arrays of nanodrops, each a mere nanoliter in volume, but holding a small cell culture sample, are being used to place hundreds of thousands of cell culture assays on a laboratory desktop, revolutionizing the speed with which new pharmaceuticals can be screened for activity. The time required for new drugs to reach patients could thus be reduced, saving human lives.

Drug delivery. Drug and gene delivery will continue to impact significantly on the practice of medicine. Nanotechnology as applied to drug delivery systems will undoubtedly dramatically improve the therapeutic potential of many water-insoluble and unstable drugs. Microsensors interfaced to a nanoscale drug delivery system could dispense precise amounts of drugs for optimum functionality and minimum toxicity. However, significant challenges still remain in synthesis and processing of drug-carrier nanoparticles at the industrial scale. Nanotechnology may also help reach the hitherto elusive goal of active drug targeting to selected cells within the body. Nanotechnology that can further reduce the size and reproducibly attach targeting ligands to the drug-loaded nanoparticles may help localize the drug to the desired tissues in the body. These nanoparticles may also be valuable tools for molecular and cell biologists to study fundamental cellular processes such as receptor-mediated endocytosis and intracellular trafficking.

Interfaces between biological and other materials. In the repair of the human body with prosthetics or artificial replacement parts, mechanical attachment to the body, or alternatively, rejection by the body, occurs at biological interfaces. The nanoscale chemical and topographical details of the implanted materials determine the reaction of the body. If we can gain sufficient understanding and control of these biological reactions to surface nanostructure, we may be able to control the rejection of artificial implants. Similarly, it may be possible to surround implanted tissue with a nanofabricated barrier that would thwart the rejection mechanisms of the host, allowing wider utilization of donated organs. Ultimately, better materials and understanding of their interaction with the body may lead to implants that the body will not only accept, but that will actually become integrated into the body. Nanofabrication and nanosynthesis give us powerful new tools to address these important medical issues for which a great deal of research is still necessary.

Various bio-inspired ideas are discussed in other chapters (e.g., Chapter 4, on synthesis, and Chapter 6, on nanodevices).

8.4 SCIENTIFIC AND TECHNOLOGICAL INFRASTRUCTURE

The infrastructure needs for nanobiology are similar to those for other fields: multiuser facilities to provide access to specialized technologies, funding mechanisms and organization structures that encourage and support multidisciplinary teams and are responsive to rapid technological change, and training of a new generation of scientists and engineers who are prepared to maximally exploit this new knowledge.

The teaming of physical scientists, engineers, biologists, and health professionals will be required for research and development efforts. The universities should be supported with grants for training new undergraduate, graduate, and postdoctoral students in these interdisciplinary areas.

8.5 R&D INVESTMENT AND IMPLEMENTATION STRATEGIES

- Fund basic science and technology development needed for future biotechnology, health, and national security (biowarfare, nanobiodevices, and survivability) needs. This must include basic research in the cell and molecular biology of the many naturally occurring nanomachines within cells.

- Fund efforts to train clinicians in the use of the emerging technologies and their integration into medical instruction.
- Promote funding in proposals with rapid turnaround times for exploratory, agile response to developing opportunities uncovered by advances in nanotechnology.
- Encourage interdisciplinary cooperation of academic, industrial, and Federal laboratories.
- Support coordinated research by teams that represent the required diversity of disciplines, at sufficient magnitude to make rapid progress.

8.6 PRIORITIES AND CONCLUSIONS

- Exploratory research should be encouraged and new ideas promoted aggressively in the area of nanobiotechnology.
- A systematic investigation should be undertaken of natural structures with intrinsic patterns at the nanoscale, as well as in use of the identified nanoscale patterns for new materials and devices.
- Interaction of biomolecules with inert materials is an area of special interest both for medical application and for understanding the role of environment on the origin and evolution of life on Earth.
- It is important to support universities in interdisciplinary training of undergraduate and graduate students at the intersection of biological, physical, and engineering sciences.

8.7 EXAMPLES OF CURRENT ACHIEVEMENTS AND PARADIGM SHIFTS

8.7.1 Special Attributes of Biological Systems

Contact person: L. Jelinski, Louisiana State University

Biological molecules and systems have a number of attributes that make them highly suitable for nanotechnology applications. For example, proteins fold into precisely defined three-dimensional shapes, and nucleic acids assemble according to well-understood rules (Figure 8.5). The ribbon diagram of the oxygen-binding protein myoglobin, found in muscle cells, is illustrated in the lower portion of the figure, a diagram constructed from atomic coordinates provided by the Protein Data Bank. Antibodies are highly specific in recognizing and binding their ligands, and biological assemblies such as molecular motors can perform transport operations. Because of these and other favorable properties, biomolecules, biophysics, and biology are themes that run through all of the topics of this report (Jelinski 1999).

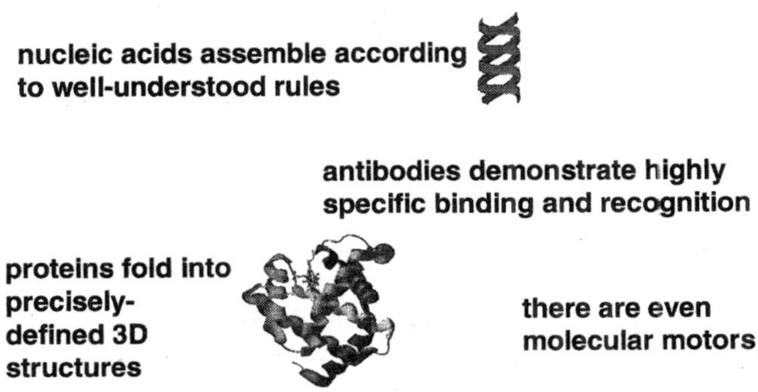

Figure 8.5. Examples of biological systems (courtesy L. Jelinski; lower diagram courtesy L. Pollack, Cornell University).

8.7.2 Nanoscience and Nanotechnology in Tissue Engineering

Contact person: D.J. Odde, University of Minnesota

Between the typical size of an animal cell, ~10 μm, and that of a protein molecule, ~5 nm, is where nanotechnology advances can effect better understanding and control of living cells. Achieving greater control of cell behavior will likely facilitate efforts in the emerging area of tissue engineering. Tissue engineering is directed toward using cells and their molecules in artificial constructs to compensate for lost or impaired body functions. Commercial ventures are currently spending ~$500 million/year in research, development, production, and marketing. In 1998 the first two tissue-engineered products came on the market after Food and Drug Administration approval (Lysaght 1998). These first two products are both engineered skin equivalents, although many more tissues are at various stages in development and clinical trials. Undoubtedly, a vast array of new nanotechnologies could potentially facilitate future tissue engineering efforts, both in basic and applied research. Four procedures are highlighted here as examples of applications of nanoscience and nanotechnology to tissue engineering.

First, scanning probe microscopy can be used to elucidate the nanometer-scale structure of protein filaments (Hameroff et al. 1990). These filaments include both intracellular and extracellular structures that are linked together via transmembrane receptors to provide the mechanical continuity that holds tissues together. Second, optical forces in the form of laser-tweezers can be used to measure motor protein motions on the nanometer scale (Svoboda et al. 1993). Understanding how molecular motors work will help us to better understand the fundamental contractile and propulsive properties of tissues. Third, biomaterials can be fabricated that have nanometer-scale features representing the imprinted features of specific proteins (Shi et al. 1999). Such imprinted surfaces could potentially provide highly stable, biospecific surfaces for the long-term maintenance of an engineered tissue equivalent. Fourth, nano/micro particles, including living animal cells, bacteria, and colloidal gold (100 nm), can be optically guided and deposited in arbitrarily defined three-dimensional arrays, a process called "laser-guided direct-writing." As shown in Figure 8.6, individual spinal cord cells can be confined and guided along a laser beam axis to generate a steady stream of particles. By combining various cell types and biomaterials, arbitrary three-dimensionally patterned cell constructs can potentially be assembled to more closely mimic the architecture and structure of native organs (Odde and Renn 1998; Renn et al. 1999).

8.7.3 Biodetection

Contact Person: J. Murday, Naval Research Laboratory

Nanotechnology promises revolutionary advances in military capability. For instance, the confluence of biology, chemistry, and physics at the nanometer scale is enabling significant advances in military sensors for biological and chemical warfare agents. Civilian disaster response teams and commercial medicine will benefit as well. We cannot afford to respond to a nerve gas attack, such as the 1995 Aum Shinrikyo incident, by carrying a canary as a sensor (Figure 8.7). Defense research and development programs are pursuing many sensor options; two related technologies are nearing fruition and will have medical applications as well.

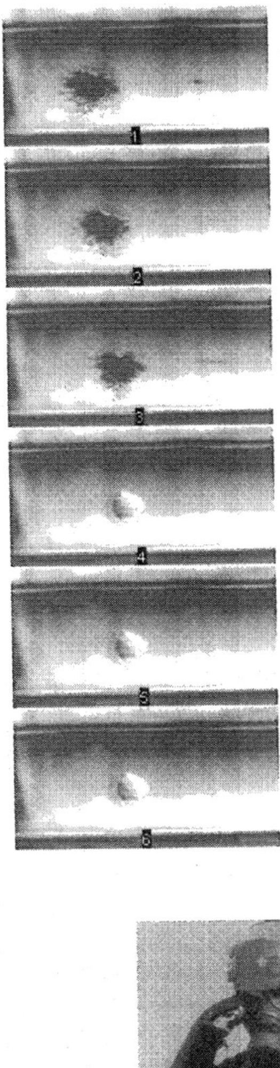

Figure 8.6. Laser-guided transport of an individual spinal cord cell inside a hollow optical fiber. The laser light comes from the left and imparts a propulsive force on the cell. The laser beam can be directed onto a surface and cells deposited into arbitrary patterns. Subcellular particles (~100-500 nm) are also guided (cell diameter, 9 μm; time interval between frames, 300 msec).

Figure 8.7. Canary sensor (courtesy Sankei Shimbun).

One is a colorimetric sensor that can selectively detect biological agent DNA; it is in commercial development with successful tests (Figure 8.8) against anthrax and tuberculosis (Mirkin 1999). Compared to present technology, the sensor is simpler, less expensive (by about a factor of 10), and more selective—it can differentiate one nucleotide mismatch in a sequence of 24, where 17 constitutes a statistically unique identification.

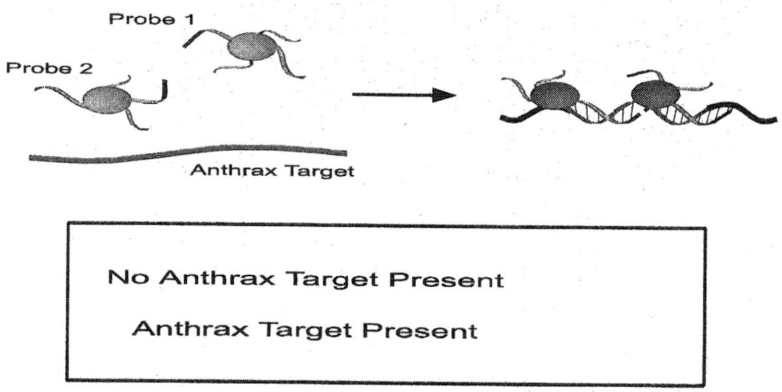

Figure 8.8. Anthrax detection: when the anthrax target is present, pairs of nanoparticles assemble together via the DNA filaments and change the color of the respective suspension (courtesy C. Mirkin, Northwestern University).

A complementary effort is based on atomic force microscopy with a sandwich immunoassay attaching magnetic beads to a microfabricated cantilever sensitive to small displacements (Figure 8.9; Colton 1999). In the laboratory this technology is already 100 to 1,000 times more sensitive than conventional immunoassays.

Both colorimetric and magnetic bead technologies might be implemented in detector arrays that provide simultaneous identification of multiple pathogens. For instance, GMR memory elements can sense the presence of the magnetic beads (Colton 1999).

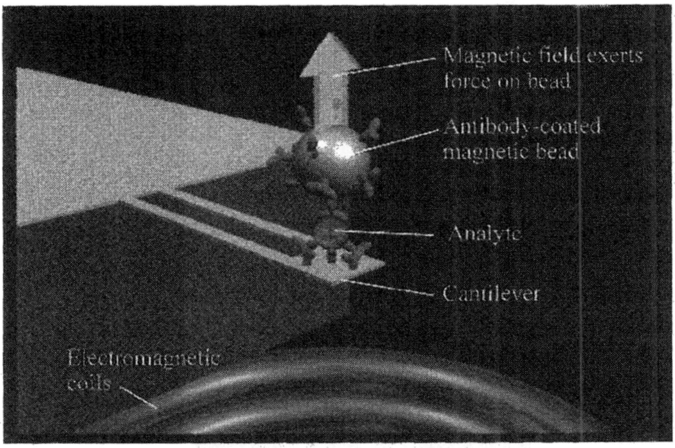

Figure 8.9. Atomic force microscope (AFM) immunoassay (courtesy Naval Research Laboratory; reprinted with permission from Baselt et al. 1996, ©1996 American Vacuum Society).

8.7.4 Semiconductor Nanocrystals as Fluorescent Biological Labels

Contact person: P. Alivisatos, University of California, Berkeley

For more than a decade there has been an intensive effort to prepare high-quality nanometer-size colloidal crystals of many common semiconductors. At the onset, this effort had a strong focus on fundamental studies of scaling laws, in this case, quantum confinement of electrons and holes. Over this decade, tremendous advances occurred in both the spectroscopy and the fabrication methods. This yielded a new class of very robust macromolecules with readily tunable emission energy. To the extent that applications of this technology were envisioned at the onset, they were focused in the domain of optoelectronics. Yet quite unexpectedly, it turns out that these colloidal nanocrystals can be used as fluorescent labels for biological tagging experiments. Biological tagging is one of the most widely employed techniques for diagnostics and visualization. As shown in Figure 8.10, it appears as though for many applications, the colloidal nanocrystals are advantageous as labels, when compared to existing organic dyes (Bruchez et al. 1998; Chan and Nie 1998). This has led to rapid commercialization of the new nanotechnology.

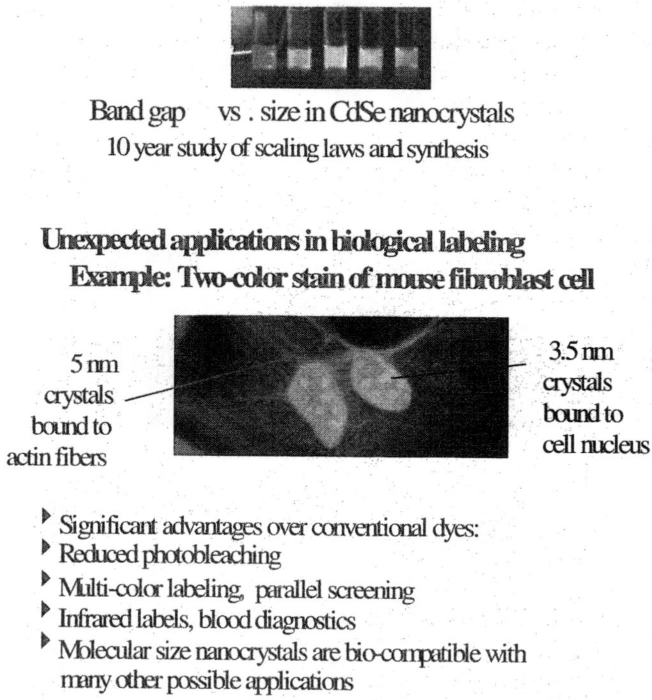

Figure 8.10. Semiconductor nanocrystals as fluorescent biological labels (reprinted with permission from Bruchez et al. 1998, ©1998 American Association for the Advancement of Science).

8.7.5 Nanofabrication of DNA "Chips"

Contact persons: M. Sussman, University of Wisconsin; P. Brown, Stanford University

DNA detector arrays that today operate in the micron size range provide the potential to do thousands of experiments simultaneously with very small amounts of material. Figure 8.11 is an image of a chip with 6,400 microdots, each containing a small amount of a different gene in the yeast genome and capable of determining how active that gene is in yeast. Yeast cells were grown under various conditions; the amount of red or yellow light represents the level of RNA produced from the DNA in that gene, under those conditions. Similar experiments using this or related technologies can now be performed with tens or hundreds of thousands of human genes. By comparing the pattern of gene expression of normal tissue with cancerous tissues, scientists can discover which few genes are being activated or inhibited during a specific disease. This information is critical to both the scientific and clinical communities in helping to discover new drugs that inhibit cancer-causing genes. The important point is that these technologies allow physiological changes in yeast or humans to be characterized, molecule by molecule, in just a few hours. Five years ago, an experiment like this would have taken dozens of scientists months to complete.

Figure 8.11. The full yeast genome on a chip (Brown 1999).

This technology therefore represents a paradigm shift in the way biologists do research, providing a means for using the vast amounts of information being revealed by the Human Genome Project. Some scientists have likened this to 150 years ago, when the periodic table for the chemical elements was discovered, ushering in a century of breakthroughs in chemistry. By analogy, the human and plant genome projects may organize all biological information in a way that may usher in a century of basic and applied research in the manipulation of life. Despite the power of the new technology, coupled with genome sequences, it is still in its nascent forms and is largely limited in its sensitivity, selectivity, and requirement for expert operators. Nanotechnology has the potential to do the following:

- Further reduce the size of the assays, allowing larger numbers of genes to be studied in each experiment
- Increase their sensitivity, for example, through better detection methods
- Result in wider application of these systems in hospitals, clinics, or perhaps even as real-time sensors within the body, for example, by enabling new ways to integrate sequential steps in lab procedures into ultraminiaturized lab-on-a-chip devices that are less subject to operator error

8.8 REFERENCES

Baselt, D.R., G.U. Lee, and R.J. Colton. 1996. Biosensor based on force microscope technology. *Journal of Vacuum Science and Technology B* 14(2):789-793.

Brown, P. 1999. http://cmgm.stanford.edu/pbrown/yeastchip.html.

Bruchez, M. Jr., M. Moronne, P. Gin, S. Weiss, and A.P. Alivisatos. 1998. Semiconductor nanocrystals as fluorescent biological labels. *Science* 281:2013-2016.

Chan, W.C.W., and S.M. Nie. 1998. Quantum dot bioconjugates for ultrasensitive nonisotopic detection. *Science* 281:2016-2018.

Colton, R. 1999. (Chemistry Division, Naval Research Laboratory -- private communication).

Jelinski, L. 1999. *Biologically related aspects of nanoparticles, nanostructured materials, and nanodevices. In Nanostructure science and technology*. NTSC Report, ed. R.W. Siegel, E. Hu, and M.C. Roco. Baltimore: World Technology Evaluation Center (WTEC). Web site http://itri.loyola.edu/nano/IWGN.Worldwide.Study/. Also published by Kluwer Academic Publishers.

Hameroff, S., et al. 1990. Scanning tunneling microscopy of cytoskeletal proteins: Microtubules and intermediate filaments. *J. Vac. Sci. and Tech.* A 8:687-691.

Lysaght, M.J. 1998. An economic survey of the emerging tissue engineering industry. *Tissue Eng.* 4:231-238.

Mirkin, C. 1999. (Department of Chemistry, Northwestern University – private communication).

Noji, H. 1998. The rotary enzyme of the cell: The rotation of F1-ATPase. *Science* 282: 1844-1845.

Odde, D.J. and M.J. Renn. 1998. Laser-based direct-write lithography of cells. *Ann. Biomed. Eng.* 26:S-141.

Renn, M.J., et al. 1999. Laser guidance and trapping of mesoscale particles in hollow-core optical fibers. *Phys. Rev. Lett.* 82:1574-1577.

Santini, J.T. Jr., M.J. Cima and R. Langer. 1999. A controlled-release chip. *Nature* 397:335-338.

Shi, H., et al. 1999. Template-imprinted nanostructured surfaces for protein recognition. *Nature* 398:593-597.

Svoboda, K. et al. 1993. Direct observation of kinesin stepping by optical trapping interferometry. *Nature* 365:721-727.

Turner, S.W., A.M. Perez, A. Lopez and H.G. Craighead. 1998. Monolithic nanofluid sieving structures for DNA manipulation. *J. Vac. Sci. Technol. B.* 16(6):3835-3840 (Nov/Dec 1998).

Chapter 9

APPLICATIONS: ENERGY AND CHEMICALS INDUSTRIES

Contact persons: D. Cox, Exxon Research and Engineering Co. (ret.); S.T. Picraux, Sandia National Laboratories

9.1 VISION

The trend to smaller and smaller structures, that is, miniaturization, is well known in the microelectronics industry, evidenced by the rapid increase in computing power through reduction of the area and volume needed per transistor on chips. In the energy and chemicals areas, this same trend towards miniaturization, i.e., control of function and/or structure at the nanoscale, also is occurring, but for different reasons. Smallness in itself is not the goal. Instead, it is the realization or now even the expectation that new properties intrinsic to nanostructures will enable breakthroughs in a multitude of different technologically important areas. Nanoengineering is expected to lead to significant improvements in solar energy conversion and storage; better energy-efficient lighting; stronger, lighter materials that will improve transportation efficiency; use of low-energy chemical pathways to break down toxic substances for remediation and restoration; and better sensors and controls to increase efficiency in manufacturing and processing.

9.2 CURRENT SCIENTIFIC AND TECHNOLOGICAL ADVANCEMENTS

What is it about the concept of nanoscale science and technology that is leading to such emphasis and excitement in a multitude of disciplines and areas of research around the world? The fundamental driving force towards efforts to exploit the nanoscale or nanostructure comes from two realizations: (1) the macroscopic bulk behavior with which we are most familiar is significantly different from quantum, or nanoscale, behavior, and (2) materials with some aspect of quantum behavior can now be synthesized and studied in the laboratory. Obviously, quantum behavior becomes increasingly important as the controlling parameter gets smaller and smaller.

Exploiting the Properties of Quantum Behavior

The following are examples where quantum behavior differs from bulk behavior:

- Nanoparticles (metal, carbon, metal oxide, etc.) exhibit novel electronic, chemical, magnetic, and structural properties.
- Molecular diffusivity through molecular sieve materials such as zeolites cannot be predicted or explained by hard sphere molecular properties or fixed wall apertures.
- Catalysts with one, two, or three dimensions in the nanometer size range exhibit unique catalytic or chemical activities.
- Electron/photon charge and transfer is different in nanostructures.

Understanding atom-to-atom interactions at the fundamental level is becoming increasingly important in materials science research efforts. There are now numerous instances where materials with some aspect of quantum behavior are actually being fabricated and studied:

- Clusters consisting of pure metals, mixed metals, metal oxides, metal carbides, metal sulfides, carbon clusters, and organic molecular clusters
- Molecular sieve materials with precisely controlled pore sizes
- Carbon nanotubes of differing helicity, diameter, and shells
- Single layer few molecular layer thin films
- Catalysts with at least one dimension at the nanoscale
- Extremely high surface area materials for use as sorbents or as catalyst supports
- Battery materials with nanoscale porosity

- Energy conversion with Gratzel cell-type devices employing dyes absorbed on nanoscale inorganic oxides
- Electronic devices where at least one device dimension is on the nanoscale

Understanding and control of nanostructure are expected to become increasingly important in many diverse areas of current and future academic and industrial activity in the chemical and energy industries, as well as in related materials endeavors:

- *In petrochemical processing*, a key goal is to promote catalytic reactions that have high selectivity with high yield. It is anticipated that this goal will be more closely approached through tailoring a catalyst particle via nanoparticle synthesis and assembly, so that it performs only specific chemical conversions, performs these at high yield, and with greater energy efficiency. Such precise control will impart opportunities for more efficient usage of our limited natural resources.
- *In energy applications,* some new nanostructured materials with well-defined pore sizes and high surface areas are currently being fabricated and tested in the laboratory for potential use in energy storage, chemical separations, and battery technologies. The use of nanoscale materials for energy generation and storage may allow for higher capacities, higher rates of charge and discharge, and far greater control over the absorption and charge transfer processes. Materials with significantly increased H_2 (or CH_4) storage capacity in small volume containers could be the enabling feature for lower-cost, more efficient, less polluting fuel-cell-powered vehicles or fuel cell power generation units for local business applications. Similarly, removal of H_2S, H_2O, CO, and/or CO_2 from natural gas near well heads would enable more efficient transport of natural gas from the well head to the end user. The relatively recent discovery that continuous channels of nanocrystalline tin aggregates are formed during the electrochemical reduction of tin oxide is opening opportunities for greatly improved rechargeable lithium-ion batteries. It has also been demonstrated that such nanostructured anode materials such as V_2O_5, $LiCoO_2$ and MnO_2 have improved capacities, lifetimes, and charge/discharge rates.

- *In the construction materials area,* novel materials are being fabricated in which improved bonding and strength dependent upon the surface area and morphology of nanoscale constituents are leading to materials with enhanced strength and toughness for use in the construction and steel industries. The use of nanostructured tungsten carbide/cobalt composites gives a two-fold increase in abrasion resistance and hardness, increasing the lifetime of drill bits. Incorporation of nanoscale carbon fibers or nanotubes into concrete not only increases strength but also offers the opportunity to continually monitor the structural integrity of the structure via electrical resistivity measurements. Proper control of the number and size of grain boundaries in steels is expected to lead to improved strength and performance. For example, smaller-diameter but stronger pipes can operate at higher pressures, which may allow more cost-effective distribution of high-pressure gases.
- *In other materials areas,* the potential is being investigated for creation of novel materials, devices, and processes, including thermal barrier materials and highly selective sensors, and also for development of molecular replication technologies for rapid scale-up and manufacturing. These are being pursued actively in laboratories worldwide.

Ultimately, of course, no matter how much is learned about the nanoscale and its properties, the industrial use of such materials will only come about if and when there is a definitive cost advantage for the end user.

The chemical and energy industries are already benefiting from the nanoscale technology revolution. Since the late 1970s, the scientific community has experienced enormous progress in the synthesis, characterization, and basic theoretical and experimental understanding of materials with nanoscale dimensions, that is, small particles, clusters, or nanocrystalline materials (Prigogine and Rice 1988; Averback et al. 1991). Such materials are groups of atoms or molecules that display properties different from both the smaller individual atoms or molecules and the larger bulk materials. Many techniques have been developed to produce clusters, beams of clusters, and clusters in a bottle (Hu and Shaw 1999, 15) for use in many different applications, as will be discussed below.

The properties of such materials have opened a third dimension to the periodic table, that is, the number of atoms (N) (Rosen 1998). N now becomes a critical parameter by which the properties for these "nanoscale" systems are defined. As a simple example, for metals, we have known for decades that the atomic ionization potential (IP) is typically about twice the value of the bulk work function. It is only relatively recently that experiments have shown that the ionization potential and electron affinity for clusters containing specific number N of (metal) atoms varies dramatically and non-monotonically with N for clusters containing less than 100-200 atoms (Taylor et al. 1992; Rohlfing et al. 1984). Other properties such as chemical reactivity, magnetic moment, polarizability, and geometric structure are also found to exhibit a strong dependence on N. The expectations for new materials with properties different from the atom or the bulk material have been realized. The opportunity is now open to precisely tailor new materials through atom-by-atom control of the composition to generate the clusters or particles of precise design for use in their own right or as building blocks in larger-scale materials or devices—that is, nanotechnology fabrication at its ultimate.

As an example of the unique properties of nanoparticles that make them of interest to energy and chemicals researchers, nanocrystalline materials composed of crystallites in the 1-10 nm size range possess very high surface-to-volume ratios, owing to their fine grain size. These materials are characterized by a very high number of low-coordination-number atoms at edge and corner sites, which can provide a large number of catalytically active sites. Nanostructured materials exhibit chemical, catalytic, and physical properties characteristic of neither the isolated atoms nor the bulk material. One of the key issues in applying such materials to industrial problems involves discovery of techniques to stabilize small nanocrystallites in the shape and size desired. This is an area of active fundamental research (Ying and Sun 1997; Trudeau and Ying 1996), and if successful on industrially interesting scales, is expected to lead to materials with novel properties specific to the size or number of atoms in the crystallite.

Nanoscale Catalysis

A key objective of nanoscale catalyst research is to produce a material with exceedingly high selectivity at high yield in the reaction product or product slate, that is, chemicals or fuels by design; with the option of altering the product or product slate simply by changing the surface functionality, elemental composition, or number of atoms in the catalyst particle. For instance, new catalysts with increasing specificity are now being fabricated in which the stoichiometry may be altered due to nanometer size restrictions in one, two or three dimensions.

Recent examples where nanocrystalline metallic and ceramic materials have been successfully investigated for catalytic applications are discussed briefly below:

- Nanostructured gold catalysts fabricated by a group at Osaka National Research Institute display novel catalytic properties (Haruta 1997a; Haruta 1997b). Bulk gold is unreactive under the same conditions. Highly selective catalytic activity at or even below room temperature is observed to switch on for gold particles smaller than about 3-5 nanometers in diameter. Accompanying this turn-on in catalytic activity is the discovery that these nanoscale gold particles (crystals) have an icosahedral structure and not the fcc structure of bulk materials. Several issues appear to be key in fabricating these novel catalytic materials. For instance, the Osaka group has shown that the preparation method is crucial to fabricating gold catalysts with high catalytic activity and selectivity; that the catalytic activity, selectivity, and temperature of operation are critically dependent on the choice of catalyst support; and that water (moisture) even in ppm levels dramatically alters the catalytic properties. Examples of novel catalytic behavior of nanoscale gold particles demonstrated to date are (a) CO oxidation at temperatures as low as -70°C, and (b) very high selectivity in partial oxidation reactions, such as near-room-temperature reduction of nitric oxide with H_2 using alumina-supported gold nanoparticles. Nanoscale gold catalysts supported on Fe_2O_3 have recently been commercialized and are being used as "odor eaters" for bathrooms in Japan.

- The importance of controlling at least one dimension on the nanoscale size range has been demonstrated for the industrially important hydrodesulfurization (HDS) reaction (Chianelli et al. 1994; Chianelli 1998). For catalysts based on the layered compound MoS_2, maximum HDS activity is obtained only on well crystallized nanosized materials, while the HDS selectivity is determined by the number of layers or "stack height" of the nanocrystalline MoS_2. In the hydrodesulfurization reaction, cyclohexylbenzene occurs only on the MoS_2 "rim" sites or those around the "edges" of the stack, whereas the pathway to biphenyl requires both "rim" and "edge" sites. Thus, the reaction selectivity is controlled by controlling the aspect ratio of MoS_2 nanoparticles. Such control of one-dimensional and two-dimensional nanostructures for selective chemical advantage is an exciting new area of research. Of course, a major industrial challenge will be to fabricate such nanocrystals in a cost-effective and commercializable form.
- Tschope et al. (1995) have demonstrated that nanocrystalline, non-stoichiometric cerium oxide (CeO_{2-x}) catalysts give rise to a substantial reduction in the temperature of selective SO_2 reduction by CO and exhibit excellent poisoning resistance against H_2O and CO_2 in the feed stream compared to that for conventional high surface area cerium oxide. Such catalysts are produced by controlled post-oxidation of cerium nanoclusters generated via inert gas condensation synthesis. The CeO_{2-x} materials were found to possess a significant concentration of Ce^{3+} and oxygen vacancies, even after high temperature (500°C) calcination.
- Electrochemical reduction of metal salts is yet another option used to control the size of nanoscale catalyst particles (Reetz et al. 1995). This has been successfully used to prepare highly dispersed metal colloids and fix the metal clusters to the substrate. Control of current density during the electrochemical synthesis process allows control of the size of the transition metal particles at the nanoscale. A combination of scanning tunneling microscopy (STM) and high-resolution transmission electron microscopy (TEM) was used to visualize surfactant molecules attached to nanostructured palladium clusters.

Hydrogen Storage

Discovery of materials with higher hydrogen storage per unit volume and weight is an active area of research in several laboratories around the world and is considered by many to be an enabling technology for vehicular fuel cell applications. Researchers at Los Alamos National Laboratory (Schwarz 1998, 93) are studying one approach that enables materials such as magnesium to be used for hydrogen storage. Magnesium is of interest because it can store about 7.7 wt% hydrogen but the adsorption/desorption kinetics are slow compared to that of metal hydrides. High surface area (nanoparticle) mixtures of Mg and Mg_2Ni are produced by ball milling of bulk materials. The addition of Mg_2Ni catalyzes the H_2 dissociation such that the rate of hydrogen adsorption increases to that comparable to $LaNi_5$ and exhibits a low pressure adsorption plateau at about 1500 torr pressure. Experiments show that the pressure plateau can be tailored through alloying. Studies with other catalysts such as FeTi and $LaNi_5$ are presently ongoing to improve both the capacity and charge/discharge rate of hydrogen storage.

A second approach to new hydrogen storage materials involves the use of nanoporous carbon fibers and carbon nanotubes. The carbon fiber materials are produced via catalytic decomposition of hydrocarbon vapors and are reported to exhibit exceptionally high hydrogen storage capacity (Baker 1998, 172). This approach also has the potential advantage of being able to reduce the total system weight, since the sorbent, in this case carbon, is (on an atom-to-atom basis) significantly lighter than a metal or metal hydride.

Rechargeable Batteries

Discovery of the ability to intercalate alkali metal ions such as lithium into a transition metal oxide crystal framework led directly to Sony's introduction of the rechargeable lithium-ion battery into the commercial market in 1991 (Nagaura and Tozawa 1990; Bubala 1997). The Li-ion battery has several advantages over the nickel-cadmium (Ni-Cd) and nickel metal hydride (Ni-MH) batteries. The Li-ion structures are extremely lightweight and compact, making them more useful for battery packs. The Li-ion battery also provides a higher nominal voltage (approximately three times that of Ni-Cd or

Ni-MH) and a higher capacity (approximately two times that of Ni-Cd or Ni-MH) for the same weight of active material (Sony 1999).

Recently, it has been discovered at Fuji that the nanostructure created in the amorphous tin oxide-based system gives significantly higher reversible capacities than the commercial systems using carbon electrodes (Brousse et al. 1998; Idota et al. 1997; Idota et al. 1995). The novelty of the tin oxide-based system is the electrochemical reduction of tin oxide to form continuous channels of nanocrystalline tin aggregates held together by the amorphous clusters of inactive oxide ions. However, the insertion and removal of lithium from the intermetallic system is accompanied by significant volume expansion (~300%), leading to disintegration and cracking of the electrode during cycling (Courtney and Dahn 1997).

Researchers at Fuji developed an ingenious solution to this problem, wherein they used a glass-forming composition primarily composed of tin, along with glass formers from group III, IV, or V of the Periodic Table. In-situ electrochemical reduction of the glass resulted in the formation of nanocrystalline tin precipitates (~7-10 nm) within the amorphous matrix phase (Li et al. 1998). The large volume fraction of the nanocrystalline tin islands enclosed by the amorphous oxide network was sufficient to form continuous channels and thereby maintain the necessary electrical conductivity within the electrode. The open structure of the glass helped to accommodate the strain associated with the volume expansion during insertion and removal of lithium from tin. The volumetric strain was also largely reduced due to the formation of nanocrystalline regions of tin.

It is believed that the nanocrystalline nature of tin precludes the formation of bulk phases of Li-Sn alloys, and the ratio of Li:Sn is therefore varied continuously, minimizing the deleterious influence of the large-volume expansion-related strain. This nanocomposite has been described by many as an active-inactive composite; the inactive matrix is the oxide glass enclosing the cluster of the active phase of tin. The realization of this concept has led to stable reversible capacities of almost twice that of carbon, and it is presently an active area of research, particularly in identification of synthetic processes and methods to stabilize the cathode structures so that the irreversible lithium loss can be eliminated.

A novel particulate sol-gel approach has been developed at Carnegie Mellon University that leads to the formation of intermixed nanophase reactants of Li_2CO_3 (less than ~20 nm) and NiO (~30-50 nm) that directly react almost instantaneously in a single step to form the desired nanocrystalline (~100-500 nm) oxide (Chang 1999; Chang et al. 1997). The nanocrystalline oxide exhibits less than 1% antisite defects. Such fine-scale perfect crystals also exhibit one of the highest first discharge capacities. Advances in materials processing and increased understanding of the materials chemistry, nanoscale structure, and electrochemical reactions will play a pivotal role in dictating the future of this prominent energy storage technology. Carbon nanotubes have been demonstrated to store Li efficiently. Their use could improve anode performance as well.

Self-Assembled Nanostructures

Chemical self-assembly of nanostructured materials such as zeolites or carbon materials occurs when a large molecular or crystalline structure results from the precise organization of a large number of molecules or atoms into a given and reproducible structure. Typically, organization through chemical self-assembly occurs because of specific interactions of the molecules or atoms among themselves with (or without) a template. The interaction of the different bonding mechanisms involved in self-assembly is an area of strong fundamental research interest. For this document, only two areas will be highlighted: zeolites and carbon materials. Both of these materials exhibit desirable characteristics of self-assembly; namely, that they are novel and reproducible structures that can be fabricated in industrially significant quantities.

Zeolitic Materials

Aluminosilicates (e.g., zeolites) are crystalline porous nanostructures with long-range crystalline order with pore sizes that can be varied from about 4 Å to 15 Å in conventional zeolites. A zeolite may exhibit a three-dimensional zeolite cage structure consisting of intersecting straight and ziz-zag channels or a simpler two-dimensional zeolite configuration. The size of the window is determined by the number of oxygens in the ring. Table 9.1 gives approximate window dimensions for zeolites as a function of the number of oxygens in the ring.

Table 9.1. Zeolite Channel "Window" Dimensions for Number of Oxygens in Ring

Number of Oxygens in Ring	Ring Diameter (Å)
4	1.2
5	2.0
6	2.8
8	4.5
10	6.3
12	8.0

As can be seen by examination of Table 9.1, molecules can pass through or be blocked from transport through or into the zeolite, depending on the size of the window of the channel in the zeolite. For example, normal hexane with a kinetic molecular diameter of about 5.1 Å can easily pass through a 10 ring or larger, whereas cyclohexane with a kinetic molecular diameter of 6.9 would pass through a 10 ring only with great difficulty. Thus, all other things being equal, a 10-ring zeolite could be used to separate mixtures of normal hexane and cyclohexane. It is this property, together with the ability to chemically modify the acidity of zeolitic materials, that makes zeolites extremely valuable as highly selective catalysts, selective sorbents, and membranes. The zeolite catalyst area alone is the basis for an industry that exceeds $30 billion annually.

In 1992, a new family of aluminosilicates (M-41S) with pore sizes between 20 and 100 Å in diameter was reported by Mobil researchers (Beck et al. 1992; Kresge et al. 1992). One of particular interest is MCM-41, which consists of hexagonal arrays of uniform 2 to 10 nanometer cylindrical pores. Not only can such materials be synthesized, but also novel structures such as "tubules-within-a-tubule" have been fabricated as mesoporous molecular sieves in MCM-41 (Lin and Mou 1996). Of particular interest is the possibility to expand the so-called "liquid crystal templating" mechanism (Huo et al. 1994; Sun and Ying 1997) to non-aluminum dopants within the silicate MCM-41 framework and to derive non-siliceous MCM-41 type of materials (Braun et al. 1996).

Another approach to synthesizing large pore and large single crystals of zeolitic materials is being pioneered by Ozin and his group at Toronto, who have demonstrated that crystals as large as 5 mm can be synthesized (Kupperman et al. 1993). The ability to synthesize such large crystals has important implications for discovery of new sensors (selective chemical adsorbents) and membrane devices (selective transport of molecular species), since large single crystals can now be available to the laboratory researcher to carry out fundamental studies of adsorption and diffusion properties with such materials.

Such materials are expected to create new opportunities for applications in the fields of separations science for use directly as molecular sieves or as new molecular sieving sorbent materials; in catalysis as heterogeneous catalysts and as supports for other catalyst materials; as well as other novel applications (Brinker 1996). The ability to synthesize zeolitic materials of precise pore size in the range between 4 Å and 100 Å continues to expand the possibilities for novel opportunities for research and technological innovation in the catalytic, separations, and sorption technologies.

Carbon Materials

The carbon-based materials considered here include fullerenes and their relatives such as endohedral fullerenes, metal-coated fullerenes, carbon nanotubes, carbon nanoparticles, and porous carbons. Since 1990 with the discovery of techniques to produce soluble carbon in a bottle, research on and with carbon materials has skyrocketed (Dresselhaus et al. 1996; Dresselhaus and Dresselhaus 1995). Not only can the molecular forms of carbon, the fullerenes, and their derivatives be synthesized, characterized, and studied for applications, but many other new carbon materials such as multi- and single-walled carbon nanotubes can now be produced in macroscopic quantities. A rich literature now exists on these new carbon materials; the following section of this report will only highlight some recent examples of particular interest in the area of high surface area materials.

Of particular interest for future catalytic applications is the recent report that not only can C_{60} be coated with metal atoms, but also the metal coating can consist of a precise number of metal atoms. For example, $C_{60}Li_{12}$ and $C_{60}Ca_{32}$ have been identified mass spectroscopically (Martin, Naher, et al. 1993; Martin, Malinowski et al. 1993; Zimmerman et al. 1995). C_{60} has been coated with a variety of different metals, including Li, Ca, Sr, Ba, V, Ta, and other transition metals. Interestingly, addition of more than 3 Ta atoms to C_{60} breaks the C_{60} cage. Replacement of one carbon atom in C_{60} by a transition metal atom such as Co or Ir is being studied for possible catalytic applications. The future technological challenge will be to discover techniques to fabricate large quantities of such materials so that we can put such catalyst materials in a bottle and not just in molecular beams.

Carbon nanotubes have the interesting property that they are predicted to be either semiconducting or conducting (metallic) depending on the chirality and diameter of the nanotube. Such materials are being studied as conductive additives to plastics and for use in electrochemical applications (Dresselhaus 1998) where the uniformity of the nanotube diameter and length is not overly critical. Another approach is to use the carbon nanotube as a template for a nanotube of an inorganic oxide. Hollow nanotubes of zirconia and yttria-stabilized zirconia have been prepared by coating treated carbon nanotubes with a zirconium compound and then burning out the carbon template (Rao et al. 1997). Finally, many groups around the world have recently demonstrated large-scale production of single-walled nanotubes, so one may anticipate a strong upsurge in the characterization and potential usage of single-walled carbon nanotubes in the future.

Porous carbons are of interest as molecular sieve materials, both as sorbents and as membranes or as nanostraws for filtration. A major research objective is to develop materials or structures with exceedingly high storage capacity per unit volume and weight for gases such as H_2 or CH_4. These could become an economic source of combustion fuel or as the means to power fuel cells for ultralow-emission vehicles or for electric power generation. Microporous hollow carbon fibers have exhibited high permeance and high selectivity as hydrogen selective membranes; development is now underway to scale up production of these membranes to commercial levels (Soffer et al. 1987; Jones and Koros 1994; Rao and Sircar 1993). As discussed earlier, carbon fiber materials produced via catalytic decomposition of hydrocarbon vapors have also recently been reported (Baker 1998, 172) to exhibit exceptionally high hydrogen adsorption capacity.

More mundane uses of nanotubes as nanometer reinforcing rods in polymers, both for structural and electrical properties, or even in concrete are being contemplated because of the "super strength" of the individual fibers (Lourie et al. 1998). Also, the substantially larger quantities of nanotubes now commercially available allows larger-scale testing of materials fabricated with nanotubes. Incorporation of conducting carbon nanotubes in construction materials such as concrete or structural plastics will allow for real-time, continual, and remote monitoring of material integrity and quality.

Thin Films

Another approach to self-assembly involves the fabrication of microporous and dense ultrathin films. Research and development of microporous thin films for use as molecular sieving membranes using inorganic crystalline materials such as zeolites or porous silica is yet another area of active research around the world. For molecular sieving membranes, one critical challenge is to discover ways to create large-scale, thin, nearly defect-free membranes. These films have potential applications in energy conversion, batteries, catalysis, and fuel cells.

One recent example of these types of thin films is the fabrication of mesoporous conducting thin films grown from liquid crystal mixtures (Attard et al. 1997). TEM reveals an ordered array of 2.5 nm diameter cylindrical holes in a 300 nm thick Pt film. The hole diameter can be varied either by changing the chain length of the surfactant molecule or by adding an alkane to the plating solution. It is interesting that this technique produces a continuous thin film with nanoscale porosity in an electrically conducting material.

Thin films of specialized coatings for corrosion, thermal, and/or chemical stability are valuable in the chemical and energy industries. The ability to coat existing reactors, tubing and other equipment exposed to harsh environments is somewhat mundane but can save on replacement capital and processing costs in both industry and research.

Novel chemical sensors may be anticipated through use of ultrathin films composed of specialized clusters. Nanoparticles provide high surface area, which allows one to detect the state of a chemical reaction, selectively and sensitively detect chemical leaks for both safety and reduction of waste, and improve the quality of the detection signals. Demonstration of the synthesis of oriented films of mesoporous silica on mica is an important step for generation of new materials (Yang et al. 1996).

9.3 GOALS FOR THE NEXT 5-10 YEARS: BARRIERS AND SOLUTIONS

Over the last two decades, the development and improvement of new techniques to fabricate and characterize nanoscale materials has fueled much of the enormous growth in nanoscale science and technology. Breakthroughs in synthesis have not only made nanoscale materials more readily available for scientific study and characterization, but also in some instances, have opened the door to large-scale industrial testing and use.

For example, atomic force microscopy and scanning tunneling microscopy are two techniques that have become major workhorses for characterization of the structural and electronic properties of nanoscale materials. The strong upsurge in interest and funding in nanoscale materials must to a large degree be credited to the recent development of these two techniques. Combined with X-ray spectroscopy, high-resolution transmission electron microscopy, and low-energy, high-resolution scanning electron microscopy, researchers now have the means to physically characterize even the smallest structures in ways impossible just a few years ago. Not only can the structure be precisely examined, but the electronic and, in some instances, the magnetic character of the nanostructures also can be mapped out.

Using the scanning probe devices, scientists not only can image individual atoms and molecules but can also manipulate and arrange them one by one in ordered arrays. Atomic manipulation to build structures is just in its infancy, but one can imagine a route to the ultimate goal of atomically tailored materials, built up atom by atom by a robotic synthesizer. The "abacus" of C_{60} molecules produced at the IBM laboratories in 1997 is an excellent example of possibilities that may lie ahead for manipulation at the atomic scale. The Angstrom Technology Partnership project in Japan is now in its second 5-year program to push the frontiers of atom manipulation closer to the commercial sector.

Of the fundamental properties that control the stability of atoms, clusters, and particles on a surface or support, particularly critical are the adsorption and adhesive energies of a metal atom or particle on a solid metal or oxide surface. An understanding of these properties that determine the stability of materials can impact applications ranging from oxide-supported metal catalysts and bimetallic catalysts to metal-ceramic interfaces used in microelectronics. With knowledge of such parameters, scientists can predict the relative strengths of the metal–metal and metal atom–support interaction energies and infer relative stabilities as a function of the composition and size of the metal cluster. Recently, it has become possible to experimentally measure the metal atom–surface bond strength on a per-atom basis using adsorption microcalorimetry on ultrathin single crystal metal or metal oxide surfaces (Stuckless et al. 1997). The technique for direct calorimetric measurement of metal adsorption energies developed at the University of Washington is based in part on earlier work first developed by King and colleagues at Cambridge University (Yeo et al. 1995). The technique, which probes interactions on an atom-by-atom or molecule-by-molecule basis, can be thought of as another "atomic probe"—one that can be expected to substantially advance our database and understanding at the ultimate nanoscale for materials: single atoms.

Similarly, new techniques are being developed to allow chemical and catalytic reactions to be followed in-situ in real time. As an example, an infrared and nuclear magnetic resonance (NMR) spectroscopic technique is being developed at the Max Planck Institute in Muelheim, Germany (Siegel et al. 1999, NTSC Report, Appendix B) to monitor kinetics of CO adsorption on 1-3 nm diameter metal colloid particles (typically Pt, Rh, or Pd) in liquids and to follow in real time the way CO organizes itself on the particles while in liquid suspension. Such techniques will allow scientists to begin to understand the metal particle properties in solution and thus infer how they might behave in real reaction mixtures. Extension of such techniques to real catalytic reactions in solution for catalyst particles of different sizes and compositions is likely in the not too distant future.

The application of combinatorial techniques for screening the multitude of reactions occurring in many chemical processes will become exceedingly valuable in the future. One can envision nanoliter reactors injecting nanoliters of reagents on nanometer catalysts arranged in an indexed array for rapid screening of hundreds if not thousands of reactions. Alternatively, arrays of nanometer catalysts selectively prepared with a known but different number of atoms (the same or different atom) could be prepared on supports and then screened for specific reactions. The use of such combinatorial techniques to determine catalytic and chemical reactions, although in its infancy, is one of the most rapidly growing areas of research interest in industry.

The area where nanoscale materials may have the greatest future impact in the energy and chemicals industries is difficult to predict, but some signs point to possibilities of substantial advancement in the areas of adsorption/separations, particularly in gas sorption and separations, and in novel chemical catalysis using nanoscale catalyst particles.

At least two major challenges must be faced before utilization and generation of nanoscale materials can become commonplace. The first is achieving *critical dimensional control* of the nanoscale structure over long times and varying conditions. In nanoscale catalyst materials, the critical chemical selectivity is likely to be intimately associated with the local environment around what is termed the "active" site. This suggests that the size, type, and geometry of the atoms making up the active site will play a critical role in defining the conditions under which this active site will be able to carry out its designed function. Fabrication of materials with exactly the same structure and composition at each active site has been and will continue to be a major challenge to materials and catalytic scientists.

The second challenge is achieving *thermal and chemical stability control* of the fabricated nanostructure. It is generally accepted that the smaller the nanostructure (active site) the more likely it becomes that the structure may move, aggregate, be poisoned, decompose, or change its shape, composition, or morphology upon exposure to thermal and/or chemical cycling. The identification of windows of stable operation in which the specific structure or material will be able to retain the desired (and designed) behavior is critical for commercial applications.

It is important to recognize that the use of nanostructuring or nanostructures to generate, fabricate, or assemble new materials, devices, or processes is at an embryonic stage. The effects of the nanostructure on materials properties and our ability to measure them will be increasingly important for future progress in development of nanotechnology-based materials for the marketplace.

9.4 SCIENTIFIC AND TECHNOLOGICAL INFRASTRUCTURE

It is important to recognize that nanoscale science and technology is not a "stand alone" field of endeavor, but rather is more of a "generic" area that is expected to have a critical impact and overlap in many areas of science and technology. The breadth of the issues covered in this document is illustrative. The fields, disciplines, and areas of expertise that fall under the nanotechnology umbrella are many and diverse. This diversity offers both an opportunity and a challenge to the scientific, technological, educational, and funding communities.

Many of the tools and instruments required for nanotechnology research are state-of-the-art, very expensive, and inaccessible to a large fraction of researchers. Progress is most apparent in this field within collaborative undertakings that combine complementary equipment and skills. A single researcher cannot do it all in this particular field. Mechanisms are needed that will provide researchers with easy access to the tools or expertise they require. One approach is to establish centers around core facilities such as universities and government laboratories. This should suffice for researchers associated with the center but does not provide close support for individual researchers or small groups located at a distance from such a center.

9.5 R&D INVESTMENT AND IMPLEMENTATION STRATEGIES

In the energy and chemical industries, long-term R&D has come under increasing pressure. Addressing the technical challenges of supporting existing technologies and improving technology in a company typically requires a disproportionate fraction of the total resources available. As a result, over the last twenty or so years, fundamental or basic research in industrial laboratories has declined significantly. The apparent expectation by many industrial R&D managers is that much of this fundamental work will be picked up by academic or national laboratory research initiatives; however, in an ever tightening financial environment, basic research has declined in those quarters as well.

If the United States is to remain the world's technology leader, mechanisms must be established first to identify critical research areas and then to ensure long-term funding for these critical research areas. One approach is to make it simpler to establish collaborative efforts between universities, national laboratories, and industry. This will not be an easy task to accomplish, because the goals of each organization are different and in some instances even conflicting; for example, academe's need to publish conflicts with industry's need to keep material proprietary.

9.6 PRIORITIES AND CONCLUSIONS

In the energy and chemical industries, nanotechnology is still at a nascent stage: significant opportunities are available and great strides forward may be anticipated. We see that even "mature" technologies such as catalysis, dispersions and coatings, microporous materials, and structural materials are already being impacted. In many industries where nanoscale approaches are critical, specific opportunities have already been identified and in many instances, the nanoscale approach to research is ongoing. Examples include the following:

- Higher selectivity catalysts through nanoscale fabrication
- Novel sorbent materials through chemical self-assembly

- High-capacity, low-volume, lightweight nanostructured materials for hydrogen and natural gas storage
- High-selectivity, high-permeance gas separations using molecular sieving membranes
- New approaches to combinatorial chemistry through nanoscale reactors and nanoscale mixing
- Improved thermal barrier coatings using nanoscale fabrication
- Strengthened construction materials using nanoscale binders
- Higher strength polymeric fibers and pipes using nanoscale fibers
- New energy conversion employing nanoscale materials for improved light gathering and higher efficiency
- Improved rechargeable batteries using nanoscale anodes and cathodes
- Improved process efficiency through the application of smart sensors

The challenges that must be faced are twofold. First, there are the scientific and technical challenges specific to nanoscale materials and approaches. Dimensional control and stability of the nanostructure or material fabricated by nanostructuring approaches must be maintained for long times and under varying conditions and environments; self-disassembly must not be allowed to occur. The efforts in this area generally encompass and depend on many different scientific disciplines. Thus, mechanisms for collaboration and interaction appear to be a necessity for substantial progress. For ultimate success, most nanotechnology requires identifiable scale-up options to produce macroscopic materials, devices, and processes.

Second, there are the challenges of allocating sufficient time and funding to build the fundamental database that will be required for new commercial technologies based upon nanoscale approaches. Since fundamental advances cannot be anticipated, basic science should not be slighted in a shortsighted approach to funding development of what we already know.

9.7 EXAMPLES OF CURRENT ACHIEVEMENTS AND PARADIGM SHIFTS

9.7.1 Towards Artificial Photosynthesis – Gratzel Cell

Contact person: D. Ginley, National Renewable Energy Laboratory

Figure 9.1 illustrates two of the approaches under investigation that could lead to systems resembling artificial photosynthesis. In photosynthesis, light is harvested by an array of collectors, and then the excited state energy is funneled to a reaction center where this excited state energy is converted into stored chemical energy. Although light harvesting is not readily attainable at present, there are a number of approaches employing composite structures that might be enabling toward this goal.

Figure 9.1a (top) illustrates work that is being done at Lawrence Berkeley Laboratories by Alivisatos (Huynh et al. 1999, Alivisatos and Katari 1995). In this case, absorbing or emitting (if you want to make an LED, you can run a solar cell in reverse) nanoparticles (CdSe) are in a conductive organic matrix. If there is an external field present between the back metallization and the transparent front contact, then when an electron/hole pair is created on the nanoparticle absorbers, they are separated and transmitted to the contacts, and current flows in the device. Conceptually, it is possible to get excited state transfer between nanoparticle absorbers.

Figure 9.1b (bottom) illustrates the Gratzel cell, as investigated by a number of laboratories (Gratzel et al. 1999). In this case, optically excited dyes inject negative charge into nanoparticle TiO_2 and photooxidize a reversible couple in the solution. Cells with this configuration have demonstrated solar conversion efficiencies of 10% and extended stability.

In both of these approaches, the nanomaterials are critical in the processes whereby light is adsorbed, charged carriers are generated and separated, and current flows in the device. Engineering the nanostructured materials and the interfaces in the devices is critical to their eventual practical application.

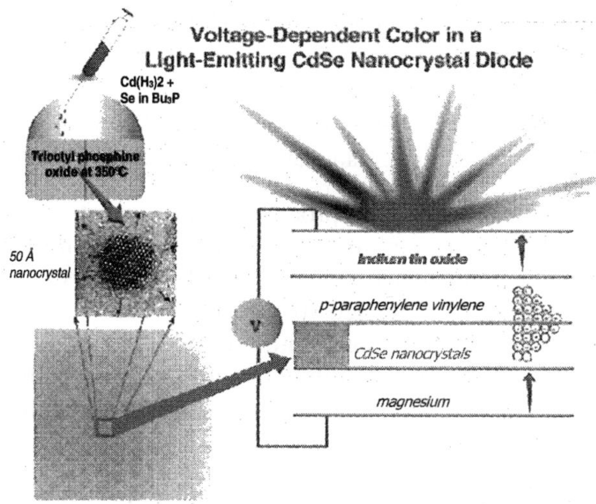

(a)

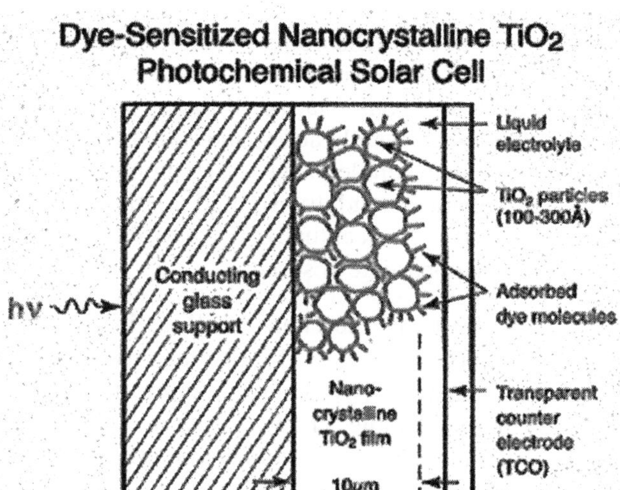

(b)

Figure 9.1. Two concepts for solar energy conversion: nanomaterials leading to direct conversion or chemical production: (a) light-emitting nanocrystal CdSe diode in a conducting organic matrix; (b) photochemical Gratzel cell of optically excited charged dyes.

9.7.2 Energy Storage: Fuel Cells, Carbon Nanotubes for Storage and Li Batteries

Contact person: D. Ginley, National Renewable Energy Laboratory

Figure 9.2 illustrates some of the new approaches under investigation for the storage of energy. These systems could be coupled to renewable energy generation systems or employed in hybrid vehicles, etc. Figure 9.2a illustrates a fuel cell that would use hydrogen and oxygen as the active couple and have only water as a product (Swartz et al. 1999). Key to the successful development of such cells is the ability to store the hydrogen effectively, to have appropriately selective membranes, and to have catalysts that readily promote the recombination of hydrogen and oxygen to produce water.

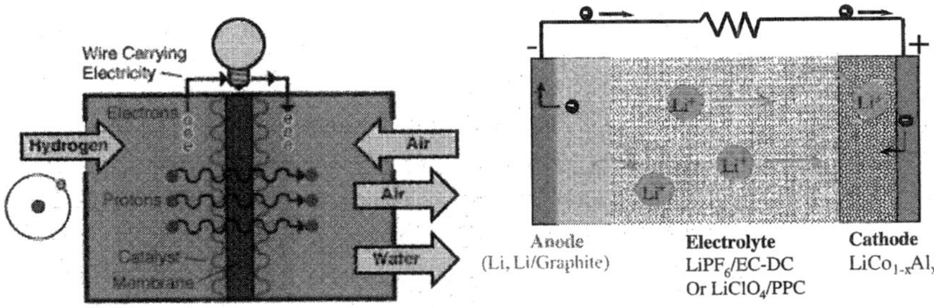

(a) (b)

Figure 9.2. Fuel cells and Li batteries for energy storage.

In addition, it is becoming clear that the membrane and catalysts will need to be nanostructured to be the most effective in a fuel cell. Figure 9.2b shows the typical structure for a rechargeable secondary Li battery. In this case, both the anode and cathode can benefit substantially from engineering at the nanoscale. Because diffusion processes are so important in these battery materials, electrodes that are mezo/nanostructured can have improved charge/discharge rates and may have greater stability (Whitehead et al. 1999). Recent work on cathodes has demonstrated that aerogels of V_2O_5 or nanoparticulate $LiCoO_2$ or MnO_2 can have improved properties as a cathode (McGraw et al. 1999). Recent work on carbon nanotubes and on Li/Sn alloys have shown that the true structure of the anode may be at the nanoscale; thus, it is critical to understand the materials science and electrochemistry (diffusion) at this length scale.

9.7.3 Improved Energy Efficiency

Contact person: S.T. Picraux, Sandia National Laboratories

Nanotechnology will have a significant influence on energy efficiency. The left side of Figure 9.3 illustrates how nanostructuring can lead to dramatic advances in magnetic materials. Here fabrication of so-called "exchange spring magnets" consisting of alternating nanometer-thick layers of hard (e.g., SmCo) and soft (e.g., Fe) magnetic materials gives dramatic increases in the performance of permanent magnets (work performed at Argonne National Laboratories). For example, a decrease in the SmCo layer thickness from 20 nm to ~2 nm for a given Fe layer thickness increases the maximum energy product $(BH)_{max}$ by over three times.

Improvements in the performance of motors due to advances in magnets will increase energy efficiency. Material structuring at the nanometer scale can also greatly increase the strength and hardness of metals and alloys, allowing the use of lighter materials (e.g., Al) and greatly reducing energy loss by friction and wear. The right side of Figure 9.3 illustrates the greatly increased yield strength of Ni by ion-implanting Al and O into the near surface (work performed at Sandia National Laboratories). The implantation produces a dispersion of nanometer-sized Al_2O_3 in the Ni, which is responsible for the increased strength. Similarly, O implantation turns Al into a material with strength and wear resistance exceeding those of the best bearing steels.

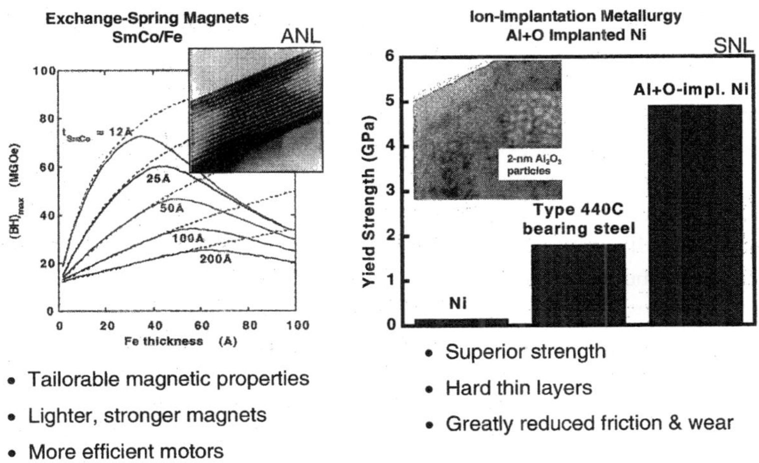

Figure 9.3. Nanoscience will lead to improved energy efficiency (DOE National Laboratories).

9.7.4 Nanoscale Catalysis

Contact person: D. Cox, Exxon Research and Engineering Co. (ret.)

The energy and chemical industries have greatly benefited from nanotechnology. Catalysis represents a major success story, both in the use of oxide-supported, highly dispersed metal (nanoscale active sites) catalysts and in the use of crystalline materials (zeolites) as highly selective catalysts. The latter case has come about most recently and is due in large part to the discovery by Mobil researchers of how to synthesize large quantities of zeolites such as ZSM5. This availability of unlimited commercial quantities of zeolites has led to a modern revolution in catalysis. Zeolite catalysts now are used to process over 7 billion barrels of petroleum and chemicals annually, forming the basis of an industry that exceeds many tens of billions of dollars a year in revenues. A multitude of commercial processes have been developed that exploit size exclusion and selective molecular diffusivity properties based upon the nanosize pore and channel structure of zeolites (Figure 9.4). This, together with the ability to tailor the acidic character of the zeolite through ion exchange processes, has opened up a myriad of research and development opportunities in the energy, chemical, and environmental industries.

One of the most important zeolite catalyzed processes in which selective molecular diffusion is the controlling aspect is the isomerization of C8 aromatics using H-ZSM5 catalysts. The demand for two C8 isomers, paraxylene and orthoxylene, is much greater than that for the C8 isomers metaxylene and ethylbenzene. H-ZSM5 with its 0.6 nm pore size has the unique ability to isomerize xylenes with little cracking of the feedstock. A second crucial property is that paraxylene has a much higher diffusivity in H-ZSM5 than do the other xylene isomers. This means that the paraxylene molecules can more easily diffuse out of the zeolite crystal, whereas the ortho and metaxylene isomers are effectively trapped within the pores.

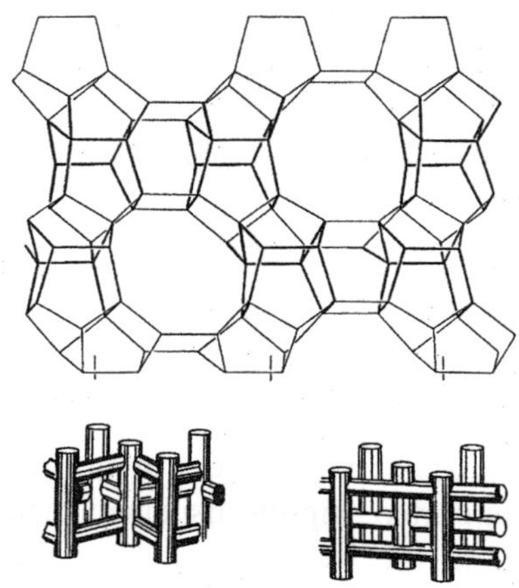

Figure 9.4. Drawing of the three dimensional channel structure of ZSM-5 (reprinted by permission from Siegel et al. 1999; ©1999 Kluwer Academic Publishers, printed version, and Loyola College in Maryland, electronic versions).

A commercial process "Parex" based upon this unique nanoscale property was developed and is in commercial use through licensing agreements in many refineries around the world. Parex is but one example of many commercial processes based upon zeolite catalytic processes used in the chemical and energy industries. Substantial industrial and academic research is ongoing to synthesize new zeolites (the recent discovery of larger pore zeolites such as MCM-41 by Mobil researchers is but one example) and develop new sorption and separation processes to exploit the molecular or nanoscale selectivity inherent in such crystals.

9.7.5 Hydrogen Storage in Carbon Nanotubes: Research Promise for Fuel Cell Applications

Contact persons: M. Dresselhaus, Massachusetts Institute of Technology, and A. Brecher, Department of Transportation, Volpe Center

The Department of Energy Hydrogen Program Plan (1992, DOE/CH10093-147) indicates that a hydrogen storage capacity of about 63 kg H_2/m^3, or 6.5 wt%, hydrogen in carbon must be achieved for advanced fuel cell power storage and generation applications. Research findings indicate that carbon nanotubes have unique structural and morphological properties that permit both a relatively large volumetric storage capacity for hydrogen gas and effective adsorption and desorption.

Carbon nanotubes are a single layer of the hexagonal lattice graphite, wrapped into a single-walled cylindrical tube of about 1 nanometer diameter (with about 20 atoms around the cylinder), but up to several microns long. This gives an aspect (or length-to-diameter) ratio in excess of 10,000. Adsorption of a hydrogen gas monolayer on a graphite surface can occur at low temperature under dense packing conditions to yield an uptake of 4.1 wt% hydrogen. H_2 molecules are considered to be spheres with a dynamic radius of 2.89 Å. Nanotubes are, however, more useful as a practical storage vessel for hydrogen, because they will retain the hydrogen even at room temperature, and because the cylinder wall curvature increases the binding energy of the hydrogen molecules to the surface carbon atoms relative to that of the free graphite surface. Furthermore, when single-wall carbon nanotubes are prepared, they form "ropes." These are a triangular, close-packed lattice of parallel cylinders with an intertube separation of 3.4 Å, the same as the separation between adjacent graphite layers. The triangular stacking of nanotubes permits additional storage of hydrogen molecules in the interstitial space.

Geometrical packing of hydrogen inside single-walled carbon nanotubes yields a storage fraction of about 3.3 wt%. Hydrogen stored in the interstitial volume would adsorb another 0.7 wt%, thus resulting in a total of 4 wt%. This theoretical prediction has indeed been experimentally confirmed by Dillon et al. as published in *Nature* in 1997 (Dillon et al. 1997) and in more recent work by the same group. Prigogine and Rice (1988) and Averback et al. (1991) discuss several materials design approaches to increasing the uptake of hydrogen and the H/C storage ratio in nanocylinders, fullerenes, or in specially designed carbon fibers. A more recent summary of presently available results (Dresselhaus and Eklund 1999) paints a promising picture for the use of carbon nanotubes for storage of hydrogen and other gases through variation of gas pressure and nanotube diameter. Efforts are being pursued to increase the hydrogen-to-carbon storage ratio to achieve the levels considered practical for multiple energy generation and storage applications, such as fuel cells for transportation and for small and medium stand-alone power sources.

9.8 REFERENCES

Alivisatos, A.P. and J.B. Katari. 1995. Optical and electrical properties of semicionductor nanocrystal assemblies. *Book of Abstracts, 210th ACS National Meeting*, Chicago, IL, August 20-24.

Attard, G.S., et al. 1997. Mesoporous platinum films from lyotropic liquid crystalline phases. *Science* 278:838.

Averback, R.S., J. Bernholc, and D.L. Nelson. 1991. In *MRS Symposium Proceedings* 206. Materials Research Society.

Baker, R.T.K. 1998. Synthesis, properties, and applications of graphite nanofibers. In *R&D status and trends in nanoparticles, nanostructured materials, and nanodevices in the United States*. Ed. R.W. Siegel, E. Hu, and M.C. Roco. Baltimore: International Technology Research Institute, World Technology (WTEC) Division, Loyola College. NTIS #PB98-117914 (also available at http://itri.loyola.edu/nano/US.Review/).

Beck, J.S., J.C. Vartuli, W.J. Roth, M.E. Leonowicz, C.T. Kresge, K.D. Schmitt, C.T.-W. Chu, D.H. Olsen, E.W. Shepard, S.B. McCullen, J.B. Higgins, and J.L. Schlenker. 1992. *J. Am. Chem. Soc.* 114:10834.

Braun, P.V., P. Osenar, and S.I. Stupp. 1996. Semiconducting superlattices templated by molecular assemblies. *Nature* 368:325.

Brinker, C.J. 1996. *Curr. Opin. Solid State Mater. Sci.* 1:798.

Brousse, T., R. Retoux, U. Herterich, and D.M. Schleich. 1998. Thin-film crystalline SnO_2-lithium electrodes. *J. Electrochem. Soc.* 145:1-4.

Brown, D.M., G. Dresselhaus, and M.S. Dresselhaus. 1997. Reversible hydrogen uptake in carbon based materials. In *MRS Conference Proc.*, Symposium Z, Dec 1-5.

Bubala, A. (Sony spokesman). 1997. Comments of April 23, posted by the battery technology team, KERI (http://lily.keri.re.kr/battery/wwwboard/).

Chang, C.C. 1999. Ph.D. dissertation. Carnegie Mellon University.

Chang, C.C., P.N. Kumta, and M.A. Sriram. 1997. Cathode materials for lithium-ion secondary cells. United States patent pending.

Chianelli, R.R. 1998. Synthesis, fundamental properties, and applications of nanocrystals, sheets, and fullerenes based on layered transition metal chalcogenides. In *R&D status and trends.*

Chianelli, R.R., M. Daage, and M.J. Ledoux. 1994. *Advances in Catalysis* 40:177.

Courtney, I.A., and J.R. Dahn. 1997. Electrochemical and in situ X-ray diffraction studies of the reaction of lithium with tin oxide composites. *J. Electrochem. Soc.* 144:2045-2052.

Dillon, A.C., K.M. Jones, T.A. Bekkedahl, C.H. Kiang, D.S. Bethune and M.J. Heben. 1997. Storage of hydrogen in single walled carbon nanotubes. *Nature* 386:377-379.

Dresselhaus, M.S. 1998. Carbon-based nanostructures. In *R&D status and trends.*

Dresselhaus, M.S., and G. Dresselhaus. 1995. *Ann. Rev. Mat. Sci.* 25:487.

Dresselhaus, M.S., and P.C. Eklund. 1999. Hydrogen adsorption in carbon materials. Preprint for MRS Bulletin (Nov.).

Dresselhaus, M.S., G. Dresselhaus, and P. Eklund. 1996. *Science of fullerenes and carbon nanotubes.* San Diego: Academic Press.

Gratzel, M.; K. Brooks, and A.J. McEvoy. 1999. Dye-sensitised nanocrystalline semiconductor photovoltaic devices. *Innovative materials in advanced energy technologies*, 577-584. Faenza, Italy: Advanced. Science and Technology.

Haruta, M. 1997a. *Catalyst Surveys of Japan* 1(61).

___. 1997b. *Catalysis Today* 36:153.

Hu, E., and D. Shaw. 1999. Synthesis and assembly. In *Nanostructure science and technology*, 1999, National Science and Technology Council (NSTC) Report, R.W. Siegel, E. Hu and M.C. Roco, Eds., Baltimore: International Technology Research Institute, World Technology (WTEC) Division. Web site: http://itri.loyola.edu/nano/IWGN.Worldwide.Study/. Also published by Kluwer Academic Publishers (1999).

Huo, Q., D.I. Margolese, U. Ciesla, P. Feng, T.E. Gier, P. Sieger, R. Leon, P.M. Petroff, F. Schuth, and G.D. Stucky. 1994. *Nature* 368:317.

Huynh, W.U., X.G. Peng, A.P. Alivisatos. 1999. CdSe nanocrystal rods/poly(3-hexylthiophene) composite photovoltaic devices. *Proc. Electrochem. Soc.* 99-11.

Idota, Y., T. Kubota, A. Matsufuji, Y. Maekawa, and T. Miyasaka. 1997. Tin-based amorphous oxide: A high-capacity lithium-ion-storage material. *Science* 276:1395-1397.

Idota, Y., M. Mishima, Y. Miyaki, T. Kubota, and T. Miyasaka. 1995. European Patent 651, 450A1.

Jones, C.W., and W.J. Koros. 1994. *Carbon* 32:1419.

Kresge, C.T., M.E. Leonowicz, W.J. Roth, J.C. Vartuli, and J.S. Beck. 1992. *Nature* 359:710.

Kupperman, A., et al. 1993. *Nature* 365:239.

Li, H., X. Huang, and L. Chen. 1998. Direct imaging of the passivating film and microstructure of nanometer-scale SnO anodes in lithium rechargeable batteries. *Electrochemical and Solid State Lett.* 1:241-243.

Lin, H.-P., and C.-Y. Mou. 1996. "Tubules-within-a-tubule" hierarchical order of mesoporous molecular sieves in MCM-41. *Science* 273:765.

Lourie, O., D.M. Cox, and H.D. Wagner. 1998. Buckling and collapse of embedded carbon nanotubes. *Phys. Rev. Lett.* 81:1638.

Martin, T.P., U. Naher, H. Schaber, and U. Zimmerman. 1993. Clusters of fullerene molecules. *Phys. Rev. Lett.* 70:3079.

Martin, T.P., N. Malinowski, U. Zimmerman, U. Naher, and H. Schaber. 1993. Metal coated fullerene molecules and clusters. *J. Chem. Phys.* 99:4210.

McGraw, J.M., C.S. Bahn, P.A. Parilla, J.D. Perkins, D.W. Readey, D.S. Ginley. 1999. Li ion diffusion measurements in V_2O_5 and $Li(Co_{1-x}Al_x)O_2$ thin-film battery cathodes. *Electrochim. Acta* 45(1-2), 187-196.

Nagaura T., and K. Tozawa. 1990. Lithium ion rechargeable battery. *Progress in Batteries and Solar Cells* Vol. 9.

Prigogine, I., and S. Rice. 1988. *Advances in chemical physics* 50, Parts 1 & 2. J. Wiley.

Rao, M.B., and S. Sircar. 1993. *Gas Separation and Purification* 7:279.

Rao, C.N.R., B.C. Satishkumar, and A. Govindaraj. 1997. *Chem. Commun.* 1581.

Reetz M.T., et al. 1995. *Science* 267:367.

Rohlfing, E.A., D.M. Cox, and A. Kaldor. 1984. Photoionization spectra and electronic structure of small iron clusters. *J. Chem. Phys.* 81:3846.

Rosen, A. 1998. A periodic table in three dimensions: A sightseeing tour in the nanometer world. In *Advances in Quantum Chemistry*, Vol. 30. February. New York: Academic Press.

Schwarz, R.B. 1998. Storage of hydrogen in powders with nanosized crystalline domains. In *R&D status and trends*.

Siegel, R.W., E. Hu, and M.C. Roco, eds. 1999. NSTC (National Science and Technology Council) Report. *Nanostructure science and technology*. Baltimore: International Technology Research Institute, World Technology (WTEC) Division. Web site: http://itri.loyola.edu/nano/IWGN.Worldwide.Study/. Also published by Kluwer Academic Publishers (1999).

Soffer, A., J.E. Koresh, and S. Saggy. 1987. United States Patent 4685940.

Sony Corp. 1999. Energy Web site, http://www.sony-media.com/en/technical/lithium-ion.html.

Stuckless, J.T., D.E. Starr, D.J. Bald, C.T. Campbell. 1997. Metal adsorption calorimetry and adhesion energies on clean single-crystal surfaces. *J. Chem. Phys.* 107:5547.

Sun, T., and J.Y Ying. 1997. Synthesis of microporous transition-metal-oxide molecular sieves by a supramolecular templating mechanism. *Nature* 389:704.

Swartz, S.L., M.M. Seabaugh, and W.J. Dawson. 1999. Nanostructured materials for electrochemical systems. *Proc. Electrochem. Soc.* 99-13.

Taylor, K.J., C.L. Pettiette-Hall, O. Cheshnovsky, and R.J. Smalley. 1992. Ultraviolet photoelectron spectra of coinage metal clusters. *J. Chem. Phys.* 96:3319.

Trudeau, M.L., and J.Y. Ying. 1996. *Nanostr. Mater.* 7:245.

Tschope, A., and J. Ying. *Nanostr. Mater.* 4:617.

Tschope, A.S., W. Liu, M. Flyzani-Stephanopoulos, and J.Y. Ying. 1995. Redox activity of nonstoichiometric cerium oxide-based nanocrystalline catalysts. *J. Catal* 157:42.

Volintine, B. 1999. Nanotechnology R&D supports DOE missions. In *IWGN Workshop Proceedings*, January 27-29 (private communication).

Yang, H., A. Kuperman, N. Coombs, S. Mamich-Afara, and G.A. Ozin. 1996. Synthesis of oriented films of mesoporous silica on mica. *Nature* 379:703.

Yeo, Y.Y., C.E. Watanaby, and D.A. King. 1995. Calorimetric measurement of the energy difference between two solid surface phases. *Science* 268:1731.

Ying, J.Y., and T. Sun. 1997. Research needs assessment on nanostructured catalysts. *J. of Electroceramics* 1(3): 219.

Whitehead, A.H., J.M. Elliott, and J.R. Owen. 1999. Nanostructured tin for use as a negative electrode material in Li-ion batteries. *J. Power Sources* 81-82, 33-38.

Zimmermann, U., N. Malinowski, A. Burkhardt, and T.P. Martin. 1995. *Carbon* 33:995.

Chapter 10

NANOSCALE PROCESSES IN THE ENVIRONMENT

Contact persons: **R. Flagan, California Institute of Technology; D.S. Ginley, National Renewable Energy Laboratory**

10.1 VISION

Nanotechnologies have the potential to significantly impact the generation and remediation of environmental problems through understanding and control of emissions from a wide range of sources, development of new "green" technologies that minimize production of undesirable by-products, and remediation of existing waste sites and polluted water sources. Removal of the finest contaminants from water supplies (less than 300 nm) and air (under 20 nm), and continuous measurement and mitigation in large areas of the environment, are envisaged.

However, nanoscale materials also potentially pose occupational and ambient health risks from both existing sources, such as diesel engines, and new systems involved in the production of nanoscale materials. In many cases, these systems are new technologies, and their environmental hazards will need to be carefully assessed.

Interdisciplinary research on molecular/nanoscale processes that take place in natural systems is important for understanding the environmental consequences of generation and transport of contaminants in the environment. Research needs include studies of interfaces between organic and inorganic structures, with focus on specific processes characterized by small length scales.

10.2 CURRENT SCIENTIFIC AND TECHNOLOGICAL ADVANCEMENTS

Nanoscale Processes in the Environment

Complex physical and chemical processes involving nanoscale structures are essential to phenomena that govern the sequestration, release, mobility, and bioavailability of nutrients and contaminants in the natural environment. Processes at the interfaces between natural physical and biological systems have relevance to health and biocomplexity issues. Increased knowledge of the dynamics of processes specific to nanoscale structures in natural systems not only will improve understanding of transport and bioavailability but also lead to development of nanotechnologies useful in preventing or mitigating environmental harm.

Pollution by Nanoparticles and Mitigation

Nanoscale materials occur naturally in the atmosphere, in minerals, in the ocean, and in biological systems. The risks to human health for particles on this length scale have not been assessed. In some cases, as for silica and asbestos fibers, the hazard potential is clear; in others, it appears that the hazard potential may be lower. Nanoscale aerosol particles are constantly involved in complex chemical processes in the atmosphere.

The considerable attention paid in recent years to the presence of fine particles in the atmosphere has led to the PM-2.5 ambient air quality standard. However, part of the difficulty of assessing the impact of nanoscale materials on biological systems is finding analytical techniques suitable for monitoring both their presence and their impact. Efforts to understand the nature of the particles at the fine end of the size spectrum have led to the development of a number of instruments that are facilitating determination of the health and environmental impacts of nanoparticles. Nanoparticles can be grown to detectable sizes by vapor condensation in an instrument called a condensation particle counter. Over the past decade, the detection capability of such instruments has been extended to as small as 3 nm in atmospheric pressure air. Additional information on the nature of nanoparticles can be obtained by differential mobility analysis, a technique in which particles are charged (typically with one positive or negative charge), caused to migrate across a particle-free stream by an applied electric field, and discharged as a monodispersed aerosol. Over the past decade, the time required to measure particles over a wide range of sizes has been reduced from tens of minutes to a fraction of a minute in commercially available instruments, and measurements within a few seconds have been demonstrated in the laboratory. As interest in nanotechnology has increased, the sizing capability of the differential mobility analysis has been extended to 1 nm and below.

Environmental Technologies

A number of environmental and energy technologies have already benefited substantially from nanotechnology.

- *Reduced waste and improved energy efficiency.* Catalysis presents a major success story. It has also been demonstrated in some cases that catalytic efficiency (rate and turnover) is substantially altered by the use of nanoscale reagents in homogeneous or heterogeneous applications. Nanoscale materials for Li battery cathodes such as aerogel or zerogel V_2O_5 can substantially increase capacity, cell life, and charge-discharge rates.
- *Environmentally benign composite structures.* The ability to incorporate nanoscale inclusions in composites has the potential to produce materials with improved properties and tailored to specific applications such as improved filtration systems. This can produce systems with increased environmental robustness, resulting in longer service life and reduced overall system costs and replacement needs, and reduced environmental impact. It also can produce lighter, smaller structures, resulting in systems with reduced energy consumption. Use of nanoscale materials in composites can range from the use of oxide- or nitride-based inclusions in steels to the development of fully engineered heterogeneous composites.
- *Waste remediation.* Nanostructured materials have an increasingly important role in the remediation of wastes. This can take many forms, from using TiO_2 particles to oxidize organic contaminates, to employing nanoscale scavengers to capture heavy metals in contaminated waste sites. In many cases, illuminated particles can be oxidizing agents, active either in solution or in

aerosols. Recent work has shown that UV-illuminated nanoscale TiO_2 can be employed to clean atmospheric contaminates, including hazardous organic chemicals, cells, and viruses, as well as to sequester hazardous chemicals. Nanoscale particles with suitable surface derivatization (ligands or reagents) can also be used to sequester heavy metals or bond to and passivate contaminated surfaces. In addition, it is clear that the more efficient chemical processes are, the less direct waste is produced. One interesting possibility is that as the surface chemistry of nanostructured materials is better understood, it will be possible to tailor the surface in nanostructured material-mediated reactions to minimize the generation of wastes.

More recently, the discovery of the ordered mesoporous material MCM-41 has extended the pore size range that is attainable in structured inorganic materials to the 10 to 100 nm size ranges. It took nearly a decade to take these materials from the initial discovery to a commercial product. One of the uses of this product is in waste remediation of heavy metals at nuclear power sites. The ability to tailor the nanostructured materials for specific ions is important both in being able to remove all the waste and segregate it, and in lowering the cost of the process.

- *Energy conversion.* Use of energy, indirectly as electricity and as fuel for transportation, is responsible for an enormous and adverse impact on the environment. Nanoscale systems offer the potential for renewable energy conversion systems with much less waste production; when this potential is coupled with improved batteries or fuel cells having nanoscale or mesoscale electrodes for energy storage in transportation, the positive impact on the environment could be tremendous. Chapter 9 of this report includes additional details on these energy-related applications.

10.3 GOALS FOR THE NEXT 5-10 YEARS: BARRIERS AND SOLUTIONS

Nanoscale Processes in the Environment

IWGN workshop participants agreed that there is an urgent need to encourage interdisciplinary research that involves novel approaches and that adopts newly developed experimental, theoretical, and computational methods for characterizing nanostructures. If interaction is increased between the community of scientists and engineers studying the fundamental properties of nanostructures and the community attempting to understand complex processes in the environment, this will hasten both communities gaining an integrated understanding of the environmental role of nanoscale phenomena. Overall, there exist opportunities in a number of areas:

- Replacement of waste-generating technologies with "green" technologies
- Improvement of process efficiency and manufacturing of smaller and lighter materials in order to reduce material and energy use
- Better understanding and control of natural phenomena and pollution through use of nanosensors and nanoelectronics

Pollution by Nanoparticles and Mitigation

The key to avoiding or remediating pollution by nanoparticles is understanding the basic science of the interaction of nanoparticles and nanostructured materials with the environment, especially with biological systems. Little is known of the key factors governing size, shape and surface chemistry. There are no effective predictive models. Before these are developed, the pollution potential of nanoparticles can only be assessed empirically, and in many cases that is quite difficult.

Nanoparticles have the potential to be "micro-reactors" that can be employed in a wide variety of circumstances to convert energy, remediate waste, and provide sensing. There are a tremendous number of uses for these microreactors. The prerequisite knowledge of the basic materials science of these systems includes fully understanding principles of synthesis, surface chemistry/derivatization, and incorporation into macro systems; as this understanding is developed, it should be possible to design microreactors for specific applications. An example of this might be use of nanoscale sensors for direct monitoring of environmental quality as well as associated process monitoring to assure optimum efficiency and minimize energy consumption of equipment. Direct environmental sensors will be important to almost all production process lines (chemical, electronic, automotive, etc.), as well as consumer use of vehicles and heating and cooling systems. As we move to smarter and smarter appliances, the need for active sensing and feedback becomes increasingly important to facilitate the "greenest" performance. In many cases, sensing can only be accomplished at the nanoscale.

Synthesis and stabilization of small nanostructures from a materials science point of view is of environmental concern. It has been demonstrated that very complex crystalline nanoparticles can be synthesized and stabilized by novel chemical and physical deposition approaches. The ability to prepare stable, isolatable particles provides the first major building block towards entirely new important technological areas: the use of nanostructured materials both as precursors to conventional and new materials and as discrete entities in larger passive and active arrays.

Environmental Technologies

Although structural nanocomposites have begun to make an impact already, over the next 5-10 years there is a tremendous opportunity to make smarter, environmentally friendly composites. The current thrust has been primarily to develop composites that have the equivalent strength at much lighter weight through incorporation of nanoparticles in polymer matrices. The objective is to nanoengineer the polymer and nanoparticles to optimize strength and weight while simultaneously adding new functionality such as chemical inertness or reactivity, conductivity, or optical properties. An example might be corrosion-resistant colored panels for automobiles.

Sorbents, membranes, and catalysts are already ubiquitous in a wide variety of systems for waste remediation, emission prevention, and "green" and energy efficiency technologies. Many active membranes, sorbents, and catalysts operate on the nanoscale. The ability to tailor these active elements at the nanoscale can significantly improve their performance and functionality. There is an increasingly critical need to proactively and continuously preserve water and air purity. Nanoscale sorbents, membranes, and catalysts have the potential to selectively target waste atoms or molecules and sequester or destroy them while not adversely affecting the remainder. Likewise, nanotechnology has the potential to create green technologies through active intervention in a wide variety of process streams, by increasing the efficiency of a process, eliminating waste, or actively treating waste as it is produced.

10.4 SCIENTIFIC AND TECHNOLOGICAL INFRASTRUCTURE

Nanotechnology spans many disciplines and suffers from the lack of common terminology and nomenclature and standards for measurement. It also suffers from researchers' lack of access to equipment. Many of the instruments and tools required for nanotechnology research are very expensive. Mechanisms are needed that will provide researchers with efficient access to the tools that they require. This is often accomplished by establishment of centers around core facilities such as universities and government laboratories, but individual investigators and small groups also need access to state-of-the-art instrumentation. By establishing mechanisms for appropriate sharing of capital-intensive facilities with researchers from other institutions, this access can be improved. Agreements to allow reasonable access to such facilities, at cost, might become an alternative to cost sharing in the acquisition of major facilities and instrumentation. For this approach to work within the typical academic laboratory, mechanisms would have to be developed for training users and controlling risk to facilities.

10.5 R&D INVESTMENT AND IMPLEMENTATION STRATEGIES

Environmental applications of nanotechnologies demonstrate their potential to revolutionize entire industries and displace major existing technologies. In spite of the few examples to the contrary, the costs and risks of transitioning from discovery to commercialization are too great and the benefits too ill-defined for large companies to undertake the wide range of development projects required to bring nanotechnologies into practice. There is a need for government-sponsored mechanisms to facilitate this development.

An increased knowledge of the dynamics of processes specific to nanoscale structures in natural systems can improve understanding of complex processes occurring in the environment and lead to development of approaches for mitigating environmental harm.

Collaboration between universities, national laboratories, and industry is also essential to the development of nanotechnology with application to the environment. This could be accomplished by establishing cross-disciplinary fellowships to support researchers for extended visits to institutions in the other sectors.

10.6 PRIORITIES AND CONCLUSIONS

Several themes should be considered priorities in developing nanoscale processes related to environmental management:

- Develop understanding and control of relevant processes, including protein precipitation and crystallization, desorption of pollutants, stability of colloidal dispersions, micelle aggregation, microbe mobility, formation and mobility of nanoparticles, and tissue-nanoparticle interaction. Emphasis should be given to processes at phase boundaries (solid-liquid, solid-gas, liquid-gas) that involve mineral and organic soil components, aerosols, biomolecules (cells, microbes), biotissues, derived components such as biofilms and membranes, and anthropogenic additions (e.g., trace and heavy metals).

- Carry out interdisciplinary research that initiates novel approaches and adopts new methods for characterizing surfaces and modeling complex systems to problems at interfaces and other nanostructures in the natural environment, including those involving biological or living systems. New technological advances such as optical traps, laser tweezers, and synchrotrons are extending examination of molecular and nanoscale processes to the single-molecule or single-cell level. Meanwhile, mathematicians are developing more effective ways to describe systems that are dynamic, multiscale, multicomponent, and exhibit emergent or aggregate behavior.

- Integrate understanding of the roles of molecular and nanoscale phenomena and behavior at the meso- and/or macroscale over a period of time. Model nanostructures should be studied, but in all cases, research must be justified by its connection to naturally occurring systems or to environmentally beneficial uses. Environments for investigation are not limited and might include terrestrial locations (e.g., acid mines), subsurface aquifers, polar environments, or the atmosphere.

10.7 EXAMPLES OF CURRENT ACHIEVEMENTS AND PARADIGM SHIFTS

10.7.1 From Discovery to Application: A Nanostructured Material (MCM-41)

Contact person: J.J. Wise, Mobil (ret.)

At a time when amorphous metal oxide supports for catalysts were well established within the chemical industry, visionary researchers within Mobil Oil Corporation undertook a long-term research program into the use of crystalline materials as catalyst supports (Figure 10.1). That program has revolutionized catalysis, displacing conventional catalyst supports for many applications (see more details in Chapter 9). In particular, the program focused on zeolites, porous materials with well-defined shapes, surface chemistry, and pore sizes smaller than 1 nm. The zeolites Y and ZSM-5 are now used widely around the world in several major catalytic processes in the petroleum and petrochemical industries, generating billions of dollars in additional revenues.

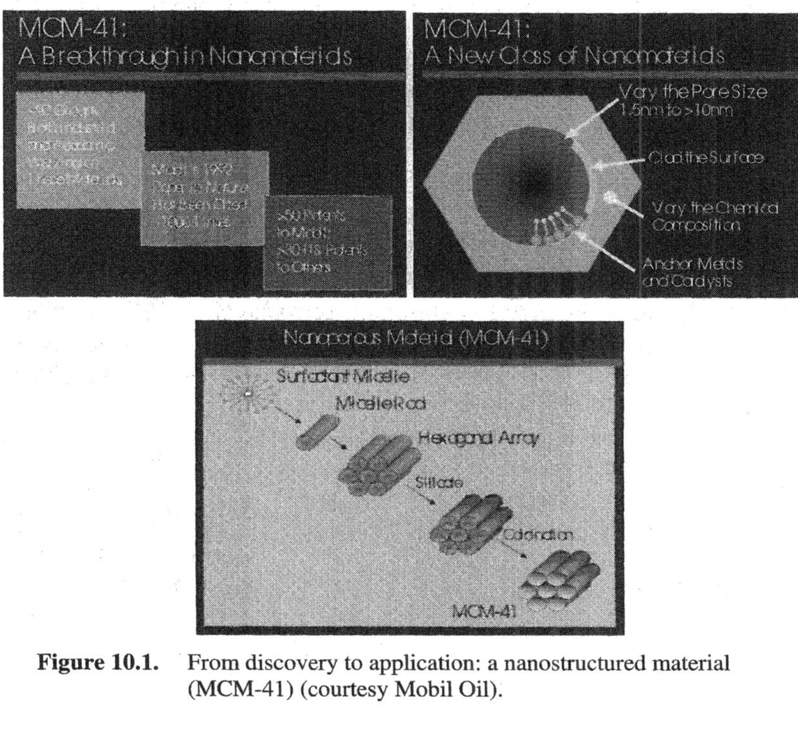

Figure 10.1. From discovery to application: a nanostructured material (MCM-41) (courtesy Mobil Oil).

10.7.2 Nanoparticles in the Environment

Contact person: A. Navrotski, University of California, Davis

Nanoparticles—iron oxides, clays, and other colloids—are the major transporting agents of both pollutants and nutrients in soil, water, and air. Understanding these modes of transport can help distribute desirable organic and inorganic constituents and immobilize undesirable ones. Control could be achieved of morphology, agglomeration, and coatings on natural nanoparticles to have the desired enhancement or retardation effect. Examples currently in use or under development are (1) zeolites and other porous soil conditioners, especially for the controlled storage and release of water; (2) use of clay and zeolites as part of radionuclide barriers at the Yucca Mountain proposed nuclear waste repository; (3) controlled release of iron, phosphorus, and other nutrients from fertilizer; (4) additions of aluminosilicates to food as texturizing agents (e.g., in non-dairy creamer) and of zeolites to animal feed (claimed to make pigs grow faster); (5) use of zeolites as ion exchangers in water purification and in detergents; and (6) use of silica gel and other nanophase solids as desiccants. Many of these are currently low-technology products; fine-tuning them may lead to more sophisticated applications, for example, drug delivery, and environmental regeneration systems in spacecraft and other confined spaces.

The ability to identify, characterize, and analyze nanoparticles quickly and in detail with respect to composition, trace elements, atomic-level structure, and morphology will lead to better understanding of their role in pollution and to better, more unambiguous, identification of the sources of pollution. For example, the definition of "asbestos," currently based solely on shape (aspect ratio >10), needs to be modified to take into account the growing body of data indicating that chemically different particles have very different toxicity. Indeed, the mechanisms of toxicity of nanoparticles are still poorly understood, and nanotechnology-based sensors are likely to play a role in developing such understanding. Similarly, nanotechnology will provide sensors for the detection of low levels of contaminants in air and water.

10.7.3 Nanoporous Polymers and their Applications in Water Purification

Contact person: D. Li and T.C. Lowe, Los Alamos National Laboratory

A completely new class of organic nanoporous polymers with narrow pore-size distribution (0.7-1.2 nm) has been synthesized using cyclodextrins as basic building blocks. These processable nanoporous polymers (Figure 10.2) exhibit superior inclusion forces and molecular transport properties towards organic guest molecules at water-solid interfaces. In fact, the formation constants of polymeric cyclodextrins and organic guest molecules are over 8 orders of magnitude larger than molecular cyclodextrins in water, and yet the process is completely reversible in organic solvents such as ethanol. The significant potential of these results is that hazardous organic contaminates may be reduced to parts-per-trillion levels in water by these polymers.

Figure 10.2. Nanoporous polymer samples in pure form (white), absorption of 4-nitrophenol (yellow), and absorption of 4-nitrothiolphenol (orange) (reprinted from Li and Ma 1999; published 1999 American Chemical Society).

10.7.4 Photocatalytic Fluid Purification

Contact person: D.S. Ginley, National Renewable Energy Laboratory

Nanoparticles have been demonstrated to have considerable environmental potential as active remediation agents. Key to the remediation of existing wastes and the prevention of new waste streams is the development of active agents that can function to remove waste where it is located or where it is generated. Figure 10.3

shows schematically how this could work and illustrates two new materials that could significantly impact this area.

Figure 10.3a illustrates how nanostructured TiO_2 can be employed as a photocatalyst to clean up a variety of waste streams (Bauer et al. 1999). The process can oxidize organic wastes and biological contaminates. It is currently being tested in operating rooms and elsewhere. This approach is illustrative of a number of approaches where a nanomaterial is inserted into a waste stream, where it reacts with or sequesters the contaminant, producing an environmentally benign process stream.

Along the lines of evolving improved materials for these applications are two new materials, illustrated in Figure 10.3b and c. In the middle of Figure 10.3b is a schematic illustration of a new inorganic fullerene composed of Mo and S (Parilla et al. 1999). This regular structure may have only relatively inert Van der Waals surfaces. With an optical bandgap in the visible region, this may be ideal species for the photooxidation of waste streams. Figure 10.3c shows a schematic of a single-wall carbon nanotube. As discussed in Chapter 9, these structures have the potential to be employed in gas purifiers and as hydrogen storage media (Kappes 1999, Dresselhaus et al. 1999). They may also have extensive use to purify air streams and to act as an agent for adsorbing heavy metals and other contaminates. Both the inorganic fullerenes and the nanotubes represent new structures that can potentially be tailored to provide specific chemical functionality.

10.7.5 Hierarchical Self-Assembled Nanostructures for Adsorption of Heavy Metals

Contact persons: G. Exarhos, Pacific Northwest National Laboratory and G. Samara and S.T. Picraux, Sandia National Laboratories

Figure 10.4 is a schematic drawing of a functionalized mesoporous nanocomposite consisting of a silicate framework of cylindrical pores that give the material a honeycomb appearance, with concomitant large surface area and nanometer porosity. The pores function as templates for the attachment of molecules of specific size and chemical functionality to form dense monolayers on the wall surfaces. The molecules bind strongly at one end to the ceramic support, leaving the free end available for interaction/reaction with targeted chemical species.

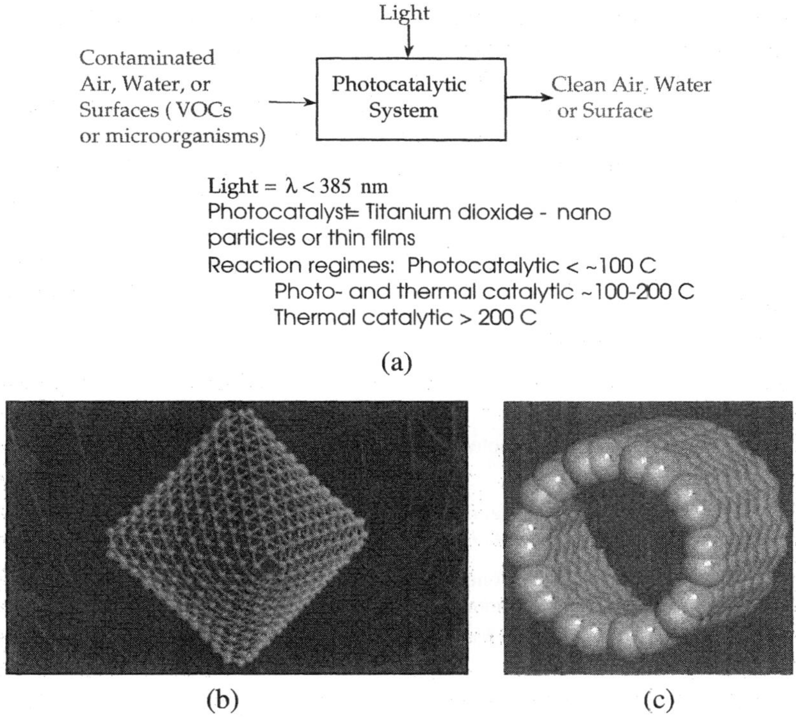

Figure 10.3. Photocatalytic fluid purification: (a) process concept; (b) inorganic MOS fullerene; and (c) single-wall carbon nanotube, each of which has photocatalytic potential.

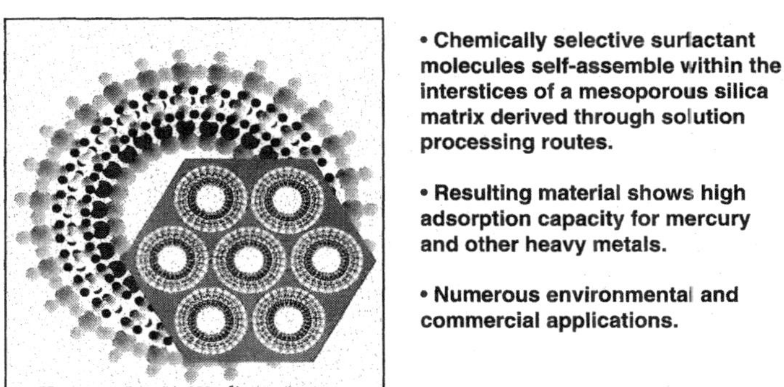

Figure 10.4. Hierarchical self-assembly for 3-D nanostructured materials: self-assembled monolayers on mesoporous supports (Pacific Northwest National Laboratory).

These nanocomposites, referred to as self-assembled monolayers on mesoporous supports, or SAMMS (work performed at Pacific Northwest National Laboratory), are very effective at sequestering heavy metal ions from waste streams and are expected to find numerous other applications in energy storage, separations, catalysis, and environmental restoration technologies.

10.8 REFERENCES

Bauer, R.; G. Waldner, H. Fallmann, S. Hager, M. Klare, T. Krutzler, S. Malato, and P. Maletzky. 1999. The photo-fenton reaction and the TiO_2/UV process for waste water treatment - novel developments. *Catal. Today* 53(1):131-144.

Dresselhaus, M.S., P.C. Eklund, and G. Dresselhaus. 1999 Fullerenes and nanotubes. *Carbon Mater. Adv. Technol.* 35-94.

Kappes, M. Carbon-based nanotechnology? 1999. *Nachr. - Forschungszent. Karlsruhe* 31(2-3):164-170.

Li, D.Q. and M. Ma. 1999. Nanosponges: From inclusion chemistry to water purifying technology. *CHEMTECH* 29(5):31-37.

Parilla, P.A., A.C. Dillon, K.M. Jones, G. Riker, D.S. Schulz, D.S. Ginley, and M.J. Heben. 1999. The first true inorganic fullerenes? *Nature* 397(6715):114.

Chapter 11

INFRASTRUCTURE NEEDS FOR R&D AND EDUCATION

Contact persons: J.L. Merz, Notre Dame University;
A. Ellis, University of Wisconsin

11.1 VISION

A substantial infusion of resources is needed for enhancement of fabrication, processing, and characterization equipment that must be made available to large numbers of users in the nanostructure community. It is also necessary to continue the process, already underway, of modifying the culture of universities to enable more interdisciplinary research to prosper, as well as to enable more industrial cooperation.

11.2 CURRENT INFRASTRUCTURE

Infrastructure for Research and Development

A major impediment to the growth of a viable nanostructure science and technology effort in the United States is an outcome of its strength: this is inherently a multidisciplinary activity. Many feel that the emphasis in this activity will shift in coming decades from the physical to the biological and life sciences. The fact that this is already happening is significant, but it is impeded by the lack of a suitable infrastructure supporting interactions among what have traditionally been very disparate disciplines. This chapter includes examples and describes in greater detail the unusual aspects of those programs that could be emulated by others to the benefit of the field overall. The infrastructure for nanoscience and technology is only in formation, and is undersized compared to the needs and overall promise of the nanotechnology field.

Education

Although change is occurring in universities in a relatively rapid fashion, there still exist many elements in the culture of our research universities that discourage multidisciplinary research. Examples include the administrative autonomy of academic departments and colleges, the fact that many centers and institutes "compete" with departments in terms of contract and grant proposal submission, the difficulties of determining (particularly with respect to tenure and promotion decisions) the relative creative contributions of faculty to multiauthored publications, and the unfortunate disconnect between research and teaching that is too often the case.

Worldwide Research Activity

In general, there appear to be two approaches to making nanostructures: (1) a so-called "top-down" approach where a nanostructure is "chiseled" out of a larger block of some material, and (2) a so-called "bottom-up" approach where nanostructures are built up from atoms and molecules using chemical techniques. The second class of nanotechnologies starts from particles, ultimately atoms or molecules, and assembles them into nanostructures.

Bottom-up nanotechnology is often called molecular engineering. It is clear that nature has been assembling atoms into complex "nanostructures" for millions of years, and in a remarkably efficient way. Molecular engineering self-assembles atoms into structures consistent with the laws of physics specified in atomic detail. The processes are also called "post-lithographic" because lithography doesn't play a central role in them (Jortner and Ratner 1998).

Bottom-up nanotechnologies have a host of important potential applications. Their impact on food production, medicine, environmental protection, even on energy production might be enormous (Gleiter 1989; Whitesides et al. 1991; Aksay et al. 1992; Drexler et al. 1993; Smalley 1995; Crandall 1996; Regis and Chimsky 1996; Freitas 1999). However, it is not enough to improve and extend the techniques of assembling molecules atom by atom: we must solve the problem of artificial self-replication and integration as well. Self-assembling atoms have been proposed and demonstrated (Smalley 1995), but no experimental verification of artificial self-reproduction has succeeded as yet. It has been shown that, in principle, self-reproducing machines in special supporting environments could be realizable but not sustainable (Von Neumann and Burks 1966, Merkle 1994). In the coming decades we shall witness the evolution of nanotechnologies at an increasing pace. U.S. National Laboratories have been devoted to this development, multidisciplinary programs have been and are being launched, and industry is contributing. The selection of both top-down and bottom-up nanofabrication tools becomes richer each year.

Nanoelectronics

Many groups are working on nanofabrication (based on semiconductors, structural and composite materials, and chemistry-based methods) and on the physical phenomena observable in these nanostructures. In the case of "nanoelectronics" (the use of nanostructures for electronic applications), research funding is shifting from the study of physical phenomena to electronic devices and circuit integration, although at present, few groups are working on the latter. It is critically important that advanced circuit architectures be developed, and these may be totally different from those used today. For example, if schemes such as quantum cellular automata (QCA) are developed (Lent 1997; Porod 1997, 1998), close collaboration with architecture design experts will be essential (Csurgay 1997). Researchers in the United States, Japan, and Europe form a very highly qualified, strong community. Their fundamental nanoscience and engineering projects are mostly funded by government sources.

In the United States, the major U.S. semiconductor companies maintain small groups (about 5-10 people) to keep informed about major developments in the area of nanoelectronics. These groups either perform basic research (e.g., Hewlett-Packard's Teramak work) or advanced development (e.g., Raytheon/Texas Instruments work on integrating resonant-tunneling devices with conventional microelectronics). The Defense Advanced Research Projects Agency (DARPA) is currently phasing out its Ultra Electronics Program, a basic research program for extremely fast and dense next-generation computing components. This was perhaps the largest U.S. Government-funded mission-oriented nanotechnology program in the United States (about $23 million/year for approximately six years). In addition, there are a few other special Department of Defense programs (e.g., MURI, URI, DURIP) that fund work in the area of nanoelectronics. The National Science Foundation also funds a few activities, including Science and Technology Centers (STCs), Engineering Research Centers (ERCs) and a recently launched project on "Partnership in Nanotechnology." All of these Government programs fund research in universities, and some (e.g., DARPA) fund programs in industry. The various university programs are listed in Section 11.7.1, and some are described in greater detail in other subsections of Section 11.7.

In Japan, most of the initiatives in the area of quantum devices and nanostructures have been funded by the Ministry of International Trade and Industry (MITI). A large part of the research work is done in industry labs (including Sony, Toshiba, Mitsubishi, NTT, Hitachi, and Motorola-Japan). Among the relatively few Japanese university groups performing advanced nanotechnology research, most notable are the University of Tokyo, Osaka University and Kyushu University. A major Japanese initiative is the R&D Association for Future Electron Devices (FED). A centralized organization manages the research, investigations, and surveys; the actual research and development work on future electron devices is subcontracted to member companies and universities; and R&D on more basic technologies is carried out by Japanese national institutes (http://www.iijnet.or.jp/fed-www/).

In Europe, ESPRIT funds two main projects as part of its Advanced Research Initiative in Microelectronics (MEL-ARI) (http://www.cordis.lu/esprit/src/melari.htm). One of these projects, OPTO, is aimed at optoelectronic interconnects for integrated circuits, and the other, NANO, at nanoscale integrated circuits. The MEL-ARI projects in general, and the NANO projects in particular, appear to mimic DARPA's Ultra Program in the United States. An ESPRIT nanoelectronics roadmap has been developed as part of the MEL-ARI initiative. The roadmap developed by the U.S. Semiconductor Industry Association (SIA) is forecasting conventional semiconductor technology, but the ESPRIT roadmap is devoted to nanoelectronics. It is known at this time that the next round of ESPRIT projects will give special attention to integration and circuit architecture of nanodevices.

11.3 GOALS FOR THE NEXT 5-10 YEARS: BARRIERS AND SOLUTIONS

There are three different levels at which the nanotechnology R&D infrastructure needs to be considered: basic research, "directed" or applied research, and development.

Basic Research

It is assumed that most of the nanoscience basic research will be done in universities and in national laboratories, because the time-line for output is too long for industry. Funding for basic research needs to be enhanced both for single investigators or small groups of faculty members, and for centers or institutes that may be located at a single campus or laboratory or involve multiple universities and national laboratories.

For individual investigators, the current size of grants is relatively small compared to the needs. It is recommended that single or principal investigator (PI) grants be increased to $200,000-300,000 per investigator, so that a PI can support several graduate students and postdocs, can purchase moderately sophisticated equipment in-house, and has the capability of accessing national equipment facilities like the National Nanofabrication Users Network (NNUN).

It is also recommended that additional centers be created of significant magnitude (on the order of $2-4 million per year per center). These centers should develop mechanisms for increasing industrial access, with personnel moving in both directions. It was noted at the IWGN workshop that there has been a tendency for successful center proposals to involve many universities, but that many of the more successful centers have been located at a single university, involving a multiplicity of disciplines crossing college boundaries. Some centers might develop new analytical or fabrication instruments, while most would focus on creating knowledge.

A useful model for further consideration in nanoscience and engineering is the Grant Opportunities for Academic Liaison with Industry (GOALI) program (http://www.nsf.gov/goali/), which funds university-industry small-group collaborative projects for fundamental research.

Directed Research

The challenges of directed or applied research in the area of nanostructures are more difficult for the single investigator model; the model of center activity is recommended as the more effective approach. Research fundamental to the integration of nanosystems is appropriate for this category. Collaboration between scientists and engineers in academe, private sector, and government laboratories needs to be integrated in the directed research programs.

Development

The development cycle for many "nanoproducts" is expected to be too long at this time for large companies and for venture capital to be able to support this research. Resources must therefore come from the Federal Government, and the work must be carried out in university and national labs and in incubators. However, to optimize the eventual commercialization of ideas generated through this research, it is essential that relationships between universities, national labs, and relevant industries be strengthened. Several recommendations are made that should encourage these relationships:

- Nanotechnology partnership programs should be formed, along the model of SBIRs, STTRs, ATP, and DARPA demonstration projects. Small high-tech companies can fill this role. Early success is apt to be in sensor and instrument areas. Grants (SBIR, etc.) can help promote the programs.
- Incubator programs should be developed at universities that support large efforts in the field of nanostructure science and technology. The university or national lab makes infrastructure available to a small company for a start-up, often with a faculty member or members taking the lead in the formation of the company. The "incubator" is a temporary intermediate stage in the formation of these start-up companies.

  For example, a technology transfer approach was adopted by the Rutgers University Center for Nanomaterials Research. This center has recognized the merit of integrating focused university research in an interdisciplinary group, with process and product development in one or more spin-off companies, each of which had its own mission, application drivers, and technical leadership. In the Rutgers University model, an organization called Strategic Materials Technologies (SMT) has been established to provide a technology development bridge between university research and industrial applications. The specifics of the SMT organization are shown in Figure 11.1. It should be noted that SMT has established several nanomaterials-focused spin-off companies, and more are in the planning stage. These small businesses have remained coupled to the university research activities.

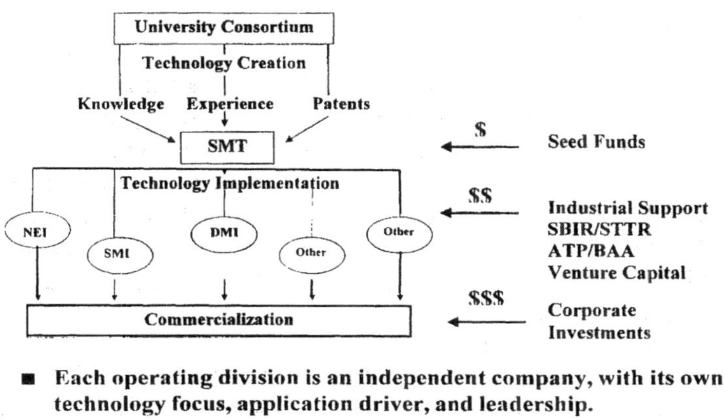

Figure 11.1. Organization and operating divisions under SMT (courtesy Nanodyne, Inc.).

One of the original groups of start-up companies, namely Nanodyne Inc., has advanced to the stage of full-scale commercialization, and hence is not shown in this diagram. The Rutgers University model should be applicable to other academic research groups and/or centers.

Finally, we note that the SMT organization is providing incentives for faculty to innovate, enabling students to gain hands-on experience in a high-tech industrial setting, and even becoming a training ground for budding entrepreneurs. In general, the level of cooperation for incubator programs should include joint submission of proposals to raise funds from state, Federal, or private sources.

A modern version of the 1870 Hatch Act that established the cooperative extension programs would be appropriate for enhancing high technology in this country. Currently, the Department of Commerce has a related program in place, the Manufacturing Extension Partnership, which should be expanded and modified to more effectively enable the development of nanotechnology.

To the extent that industry pays the cost of university research, the issue of intellectual property rights needs further discussion and investigation, because this represents a significant barrier to the development of strong industry-university relationships. It is essential that universities and industry work together to understand their mutual problems and develop solutions that encourage the transfer of technology to their mutual benefit. The Semiconductor Research Corporation has considerable experience in the area of intellectual property, which may be useful to other companies and industry consortia. Models at CalTech and Rutgers should be reviewed.

11.4 SCIENTIFIC AND TECHNOLOGICAL INFRASTRUCTURE

Infrastructure for Research and Development

National equipment user facilities such as NNUN (see Section 11.7.4) offer a partial model solution to the infrastructure problems described in Section 11.2. It is essential that a network of inexpensive and "user-friendly" user facilities be established that brings together the strategic components of activities in the physical sciences (microelectronic technologies such as CMOS and III-V semiconductor optoelectronics, organic and polymer materials, MEMS, displays, etc.) with the fundamental research activities in biology. These labs should be modular and flexible, staffed by professionals, and located where they are easily accessible to university, industry, and national lab users at a reasonable cost. Existing NNUN sites must broaden their capabilities, and the NNUN model must be extended to open many existing labs to outside users who are presently excluded. The NNUN charter already contemplates this, but funding must be provided to defray the additional costs of servicing outside users at the newly "opened" laboratories.

In addition to processing and fabrication capabilities, laboratories in research centers must make available characterization and measurement capabilities at the leading edge. For example, a central national facility having state-of-the-art scanning probe techniques of use in physics, engineering, and biology research activities should be part of this network.

In addition to the NNUN model, nanotechnology development will require a prototype fabrication facility located at a national laboratory such as Sandia National Laboratories, or at a company. This facility must be modular to accommodate MEMS, optical, chemical, and biological systems, in parallel with modern microelectronics technologies such as CMOS.

Encouragement of long-term nanotechnology and nanoscience R&D in industry is highly desirable. Infrastructure development for both start-up companies and existing companies should be stimulated by policies that facilitate a long-term focus. Furthermore, policies and funding programs should be initiated to ensure that, wherever possible and appropriate, there is sharing of nanoscience and technology R&D facilities among universities, government laboratories, and industry. In addition, as discussed in Section 11.3, there should be an emphasis on fellowships, traineeships, and internships to encourage cross-pollination of ideas among these three R&D sectors.

Education

It is not clear that the educational problems described briefly in Section 11.2 represent an area for Federal Government action, but programs could be devised that provide incentives to correct these situations and could be incorporated into solicitations for funding.

Modes of support. It must be emphasized that support of research at all levels, including single investigators, small groups, and large centers and institutes, is essential; nevertheless, multidisciplinary research centers address some of the specific issues necessary for technological education. They provide both horizontal and vertical integration of education, with students at all levels of their training interacting: undergraduate and graduate students, postdocs, and junior and senior faculty. Involvement in a research center provides a student with breadth of experience in an environment where researchers in related fields interact and work on common problems, while the student's own research still provides the necessary depth of experience.

Outreach. To generate and maintain public support for nanostructure science and technology, significant outreach activities must be undertaken. These activities must involve students at all levels (college and pre-college), and should include a general effort to popularize this research. The NSF emphasis on educational outreach is producing significant educational benefits in current NSF centers.

Curriculum development. Curriculum development is most important to enable interdisciplinary training, particularly if a marriage between the physical and biological sciences is to become a reality. One approach might be to make some form of curriculum development a requirement for research center funding.

As an example of possible curricular changes, consider electrical engineering (EE). EE departments should be aware of the potentials, and should contribute to the development, of design-fabrication and test techniques for nanoelectronics. Choices to be made include the following:

1. Should students be allowed to major in design, fabrication, and test, or only in design and test?
2. If fabrication is not included, should cooperation with a "foundry" be recommended?
3. If fabrication is included, is top-down or bottom-up technology preferred?
4. Depending on the availability of fabrication technology, either the QMOS (quantum metal oxide semiconductor) or QCA (quantum cellular automata) approach could be chosen. It is recommended that the construction of a CAD-T (computer-aided design and test) system be established for circuit level design and testing of nanoelectronic devices and circuits.
5. The construction of CAD-T systems should involve parallel research activities in several areas: device modeling, dynamic circuit simulation, device and interconnection (layout) design, and device and circuit characterization.

11.5 R&D INVESTMENT AND IMPLEMENTATION STRATEGIES

Nanotechnology R&D requires a balanced, predictable, strong, but flexible infrastructure to stimulate the further rapid growth of the field. Ideas, concepts, and techniques are moving at such an exceedingly rapid pace that the field needs coordination and focus from a national perspective. Demands are high, and the potential is great for universities and government to continue to evolve and transition this science and technology to bring forth the technological changes that will enable U.S. industry to commercialize many new products in all sectors of the economy. Even greater demands are on industry to attract new ideas, protect intellectual property, and develop appropriate products.

Tools must be provided to investigators in nanotechnology for them to carry out state-of-the-art research to achieve this potential and remain competitive. Centers with multiple grantees or laboratories where these tools would be available for this support should be established at a funding level of several million dollars annually. In addition to university- and government-led centers and networks, co-funding should be made available to industry-led consortia that will provide a degree of technology focus and different areas of relevance that are not always present in academic-led consortia. These centers should also have diverse research teams that will be effective in different scientific disciplines. Funding is needed for supporting staff to service outside users at existing and new centers. We should also investigate means to achieve the remote use of these facilities. Funding mechanisms that encourage centers and university-national laboratory-industrial collaboration should be emphasized, as well as single investigators who are tied into these networks.

Support to single investigators should provide a corresponding level of personnel and equipment support. University grants should encourage work among research groups to make maximum use of concepts and ideas being developed in other disciplines. The infrastructure must include building links between researchers, developers, and users of nanotechnology innovations. The focus must be on developing critical enabling technologies that will have significant value added in many industries.

It will also be necessary to fund training of students and support of postdocs under fellowships that will attract some of the best students available. Students should receive multidisciplinary training in various nanotechnology fields. Both organizational attention and funding should also be devoted to ensuring the open exchange of information in multidisciplinary meetings and rapid publication of results through, for example, workshops and widely disseminated summaries of research.

11.6 PRIORITIES AND CONCLUSIONS

Because of the fundamental and highly interdisciplinary nature of research and development of nanostructures, a broad and balanced approach to research funding and user facilities should be established in this country. It is recommended that nanostructure research be given the highest priority for a Federal high technology funding initiative.

Fostering collaboration between scientists and engineers in academe, private sector, and government laboratories is a priority. Multidisciplinary R&D partnerships through programs such as SBIR, STTR, ATP, and DARPA, as well as incubators at universities, should be encouraged. Industry and universities should be encouraged to participate in a review of intellectual property rights issues.

Funding of fellowships, traineeships, and internships not tied to one discipline at all levels—from high school and college students through senior investigators—is necessary to ensure free flow of ideas among disciplines, areas of relevance, and R&D sectors.

11.7 PRESENT U.S. NANOTECHNOLOGY EFFORTS

Included below are a number of examples of ongoing programs in the U.S. Government and university laboratories and in private industry. Some are merely noted; others are described in greater detail. The opinions expressed to the IWGN were diverse, as is illustrated by the statements presented below. Despite the successes highlighted here, a number of weaknesses in the U.S infrastructure, such as insufficient measuring and fabrication equipment for R&D and smaller efforts in areas such as nanodevices and ultraprecision engineering, may put the United States behind in the international effort to harness the discoveries. The centers and facilities outlined below exemplify some successful models for further development of the field. Section 11.7.2 lists a number of published accounts of cutting-edge nanoscience research selected by the authors of this chapter.

11.7.1 Federal, Industry, and University Research Programs on Nanoscience, Engineering, and Technology in the United States (selected by the chapter authors)

Contact persons: J.L. Merz, Notre Dame University, and A. Ellis, University of Wisconsin (for additional references see Siegel et al. 1999, NSTC Report)

Federal and Industry Research Programs

California Molecular Electronics Corporation (CALMEC): Molecular Electronics

Defense Advanced Research Projects Agency (DARPA): The Ultra Electronics Program (http://www.darpa.mil/mto/ultra/index.html)

Foresight Institute: Nanotechnology (http://www.foresight.org); (http://www.nanothinc.com/)

Hewlett Packard Lab: Teramak Program

IBM: Nanotech program (http://www.almaden.ibm.com/vis/vis_lab.html), with its corresponding laboratory abroad, Zurich Research Laboratory, where research is underway on microscopy at the atomic level

MITRE Corporation: Nanoelectronics and nanocomputing (http://www.mitre.org/technology/nanotech)

Molecular Manufacturing Enterprises, Inc. (MMEI)

Molecular Nanotechnology NanoLogic, Inc.: Integration of nanotechnology into computers

Nanophase Technologies Corporation

NanoPowders Industries (NPI)

Nanotechnology Development Corporation

NASA: Nanotechnology, nanoelectronics (http://www.ipt.arc.nasa.gov)

National Institute of Standards and Technology (NIST): Nanostructure fabrication

Naval Research Laboratory (NRL): Nanoelectronics processing facility

National Science Foundation (NSF): Partnership in Nanotechnology; Nanoscale processes in biological systems (updated programs on the Web site: http://www.nsf.gov/nano)

Office of Naval Research (ONR): Nanotechnology, nanoelectronics

Raytheon Co.: Nanoelectronics

Texas Instruments: projects on QMOS program and TSRAM (tunneling-based static RAM)

Xerox Palo Alto Research Center (PARC): Nanotechnology, molecular nanotechnology (http://nano.xerox.com/nano)

Zyvex Co.: Molecular manufacturing

Universities

Arizona State University: Nanostructure Research Group

CalTech: Materials and Process Simulation Center (http://www.theory.caltech.edu/~quic/index.html)

Cornell University: Cornell Nanofabrication Facility (http://www.nnf.cornell.edu); NSF Science and Technology Center for Nanobiotechnology (http://www.research.cornell.edu/nanobiotech/)

Georgia Institute of Technology: Nanocrystal Research Laboratory; nanostructure optoelectronics

Johns Hopkins University: Center for Nanostructured Materials (http://www.pha.jhu.edu/groups/mrsec/main.html)

Massachusetts Institute of Technology: NanoStructures Laboratory (http://www-mtl.mit.edu/MTL/NSL.html)

National User Facilities (NSF sponsored) in x-ray synchrotron radiation, neutron scattering, and high magnetic fields provide access to major facilities for the benefit of researchers in a wide range of science and engineering fields including nanoscience and engineering (http://www.nsf.gov/mps/dmr/natfacil.htm)

New Jersey Institute of Technology: Nonlinear Nanostructures Laboratory (NNL)

NNUN, a partnership involving NSF and five universities (Cornell University, Stanford University, UC Santa Barbara, Penn State

University and Howard University) (see Section 11.7.4 below and http://www.nnun.org/)

Oxford Nanotechnology (MA): Molecular nanotechnology, nanolithography

Pennsylvania State University: Nanotechnology

Princeton University: Nanostructure Laboratory

Rensselaer Polytechnic Institute: Nanolab

Rice University: Center for Nanoscale Science and Technology (fullerenes)

Stanford University: Stanford National Nanofabrication Users Network (NNUN) (http://snf.stanford.edu/NNUN); (http://feynman.stanford.edu/qcomp)

University of California, Santa Barbara: NSF Science and Technology Center for Quantized Electronic Structures (QUEST) (http://www.quest.ucsb.edu)

University of Illinois at Urbana-Champaign: Beckman Institute (http://130.126.116.205/research/menhome.html); Molecular and Electronic Nanustructures Group

University of Notre Dame: Center for Nanoscience and Technology

University of Washington: Center for Nanotechnology

University of Wisconsin at Madison: Center for Nanostructured Materials and Interfaces (http://mrsec.wisc.edu)

Washington State University: Nanotechnology Think Tank

Yale University: Optoelectronic structures/nanotechnology

11.7.2 Sources of Information on Nanostructures (selected by the chapter authors)

Contact person: J.L. Merz, Notre Dame University

Amlani, I., A.O. Orlov, G. Toth, C.S. Lent, G.H. Bernstein, and G.L. Snider. 1999. Digital logic gate using quantum-dot cellular automata. *Science* 284:289-91.

Ando, T., et al., eds. 1998. *Mesoscopic physics and electronics.* Berlin: Springer Verlag.

Asai, S., and Y. Wada. 1997. Technology challenges for integration near and below 0.1 μm. In *Proc. IEEE.* 85 (4):505-520, April.

Ashoori, R.C. 1996. Electrons in artificial atoms. *Nature* 379:413-417.

Beth, T. 1997. Quantum computers—a new concept in nanoelectronics. In *Proceedings of the ECCTD'97*, Budapest, 271.

Chen, M., and W. Porod. 1995. Design of gate-confined quantum-dot structures in the few-electron regime. *J. Appl. Phys.* 78:1050-1057.

Chua, L.O. 1997. CNN: A vision on complexity. *Int. Journal of Bifurcation and Chaos* 7(10): 2219-2425.

Chua, L.O., and A.T. Roska. 1993. The CNN paradigm. *IEEE Trans. Circuits Syst.* 40(3): 147-156, March.

Csurgay, A.I. 1997. The circuit paradigm in nanoelectronics. In *Proceedings of the ECCTD'97*, Budapest, 240-246.

Csurgay, A.I., W. Porod, and C.S. Lent. 1998. Signal processing with next-neighbor-coupled time-varying quantum-dot arrays. PHASDOM '98, Neuchatel, Switzerland.

Datta, S. 1989. *Quantum phenomena.* Reading, Mass: Addison-Wesley.

Deutsch, D. 1985. Quantum theory, the Church-Turing principle and the universal quantum computer. In *Proc. R. Soc. London* A400:97-117.

Drexler, E.K. 1992. *Nanosystems: Molecular machinery, manufacturing and computation.* New York: John Wiley and Sons.

Drexler, E.K., C.H. Peterson, and G. Pergamit. 1993. *Unbounding the future: The nanotechnology revolution.* New York: Quill Books.

Ellenbogen, J.C., et al. 1997. Review of nanoelectronic devices. In *Proc. IEEE.* 85(4): 521-540, April.

Ferry, D. and S.M. Goodnick. 1997. *Transport in nanostructures.* Cambridge, U.K.: Cambridge University Press.

Feynman, R.P. 1961. There is plenty of room at the bottom. In *Miniaturization.* New York: Reinhold.

Freitas, R.A. 1999 (in press). *Nanomedicine.*

Gershenfeld, N., and I.L. Chuang. 1998. Quantum computing with molecules. *Scientific American* (June): 66-71.

Grabert, H., and M.H. Devoret. 1992. Single-charge tunneling Coulomb blockade phenomena in nanostructures. *NATO ASI Series, B* 294. Plenum Press.

Hess, K., and G.J. Iafrate. 1992. Approaching the quantum limit. *IEEE Spectrum* 44-49.

Iafrate, G.J., and M.A. Stroscio. 1996. Application of quantum-based devices: Trends and challenges. *IEEE Trans. on Electron Devices* 43(10): 1621-1625.

Jortner, J. and M. Ratner, eds. 1997. *Molecular electronics: A 'chemistry for the 21st century' monograph.* Oxford, U.K.: Blackwell Science Ltd.

Kouwenhoven, L. 1995. Coupled quantum-dots as artificial molecules. *Science* 268: 1440-1441.

Lent, C.S. 1997. Dynamics of quantum-dot cellular automata and cellular nonlinear networks. In *Proceedings ECCTD '97*, Budapest, 254-258.

Lent, C.S., and P.D. Tougaw. 1997. A device architecture for computing with quantum-dots. In *Proc. IEEE* 85(4): 541-547.

Lent, C.S., P.D. Tougaw, W. Porod, and G.H. Bernstein. 1993. Quantum cellular automata. *Nanotechnology* 4:49-57.

Lent, C.S., W. Porod, G. Bernstein, and J.H. Luscombe. 1993. Current issues in nanoelectronic modeling. *Nanotechnology* 4:21-40.

Mahler, G., and V.A. Weberruss. 1995. *Quantum networks—dynamics of open nanostructures.* Berlin: Springer-Verlag.

Mazumder, P., et al. 1998. Digital circuit applications of resonant tunneling devices. In *Proceedings of the IEEE* 86(4): 664-688.

Merkle, R.C. 1998. Making smaller, faster, cheaper computers. In *Proceedings of the IEEE* 86(11): 2384-2386.

Merkle, R.P. 1994. Self-reproducing systems and low cost manufacturing. In *The ultimate limits of fabrication and measurement.* Ed. M.E. Welland et al. Dordrecht: Kluwer.

Nanotechnology database (http://itri.loyola.edu/nanobase/). Baltimore: International Technology Research Institute, World Technology (WTEC) Division.

NANOTHINC. 1997. Introduction to NanoWorld, NanoScience and NanoMarkets (http://www.nanothinc.com).

Nelson, M., and C. Shipbaugh. 1995. *The potential of molecular manufacturing*. Santa Monica: RAND Corporation. ISBN: 0833022873 (http://www.rand.org).

Orlov, A.O., I. Amlani, G.H. Bernstein, C.S. Lent, and G.L. Snider. 1997. Realization of a functional cell for quantum-dot cellular automata. *Science* 277:928-930.

Pease, R.W.F., ed. 1991. Special issue on nanoelectronics. *Proc. IEEE* 79(8).

Porod, W. 1997. Quantum-dot devices and quantum-dot cellular automata. *Int. J. of Bifurcation and Chaos* 7.

Porod, W., ed. 1998. Special issue on computational electronics. Papers presented at the Fifth International Workshop on Computational Electronics, Notre Dame, *VLSI Design* 8(1-4).

Preskill, J. 1997. Quantum computing: Pro and con. Caltech, CALT-Quic-97-031.

Crandall, B.C., ed. 1996. *Nanotechnology: Molecular speculations on global abundance.* Cambridge, Mass.: MIT Press.

Regis, E., and M. Chimsky. 1996. *Nano: The emerging science of nanotechnology*. Little Brown and Co.

Smalley, R.E. 1995. Nanotechnology and the next 50 years. Presentation to the University of Dallas Board of Councilors, December 7, 1995. http://cnst.rice.edu /dallas12-96.html.

SIA (Semiconductor Industry Association). 1997. *The national technology roadmap for semiconductors (NTRS).* San Jose, California: SRI.

Sun, J.P., G.I. Haddad, P. Mazumder, and J.N. Schulman. 1998. Resonant tunneling diodes. In *Proceedings of the IEEE* 86(4) 4:641-663.

Taur, Y., et al. 1997. CMOS scaling into the nanometer regime. In *Proc. IEEE* 85(4): 505-520.

Turton, R. 1995. *The quantum dot: A journey into the future of microelectronics*. Oxford University Press.

Von Neumann, J., and A.W. Burks. 1966. *Theory of self-reproducing automata*. University of Illinois Press.

Williams, C.P., and S.H. Clearwater. 1997. *Explorations in quantum computing*. Springer.

11.7.3 Samples of Courses on Nanoscale Science and Engineering Offered in U.S. Universities

Contact persons: A. Ellis, University of Wisconsin, Madison, and M.C. Roco, National Science Foundation

Advanced quantum devices, University of Notre Dame, EE 666

Nano-course, Cornell Nanofabrication Facility (A. Clark, M. Isaacson)

New technologies, University of Wisconsin, Madison (R. Hamers)

Nanostructured materials, Rensselaer Polytechnic Institute (R.W. Siegel, P.M. Ajayan)

Colloid chemical approach to construction of nanoparticles and nanostructured materials, Clarkson University (J.N. Fendler)

Nanoparticles processes, Yale University (D. Rosner)

Nanorobotics, University of Southern California (A. Requicha)

Nanotechnology, Virginia Commonwealth University (M. El-Shall)

Chemistry and physics of nanomaterials, University of Washington (Y. Xia)

Scanning probes and nanostructure characterization, Clemson University (D. Correll)

Nano-scale physics, Clemson University (D. Correll)

11.7.4 NNUN Network

Contact person: J. Plummer, Stanford University

Figures 11.2-11.5 outline the basic facts about the National Nanofabrication User's Network (NNUN), established in 1994 by NSF at Cornell University, Stanford University, Penn State University, University of California in Santa Barbara, and Howard University.

National Nanofabrication Users Network:
What is it ?

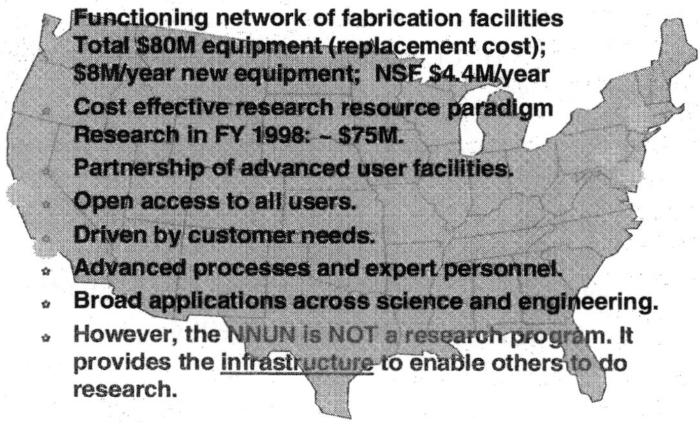

Figure 11.2. National Nanofabrication Users Network.

NNUN:
Network Vision

- **Productive "sand box" for**
  - **new approaches to nanofabrication**
  - **new applications of nanofabrication**
- **Education:**
  - **training**
  - **disseminate results**
  - **technology transfer**
- **Responsive to user needs.**
- **Sensitive to new and developing areas.**
- **Maintain leading edge nanotechnology.**
- **Catalyze new developments in nanotechnology.**

Figure 11.3. NUNN: Network Vision.

NNUN: What Does it Provide?

- **Lithography-based μm & nm-scale science and technology.**
- **State-of-the-art facilities, equipment, & processes.**
- **Enables state-of-the-art research in broad areas.**
- **Efficient use of expensive resources.**
- **Critical mass:**
  - **equipment, personnel, facilities, fabrication expertise**
- **Outreach to new users and new disciplines.**

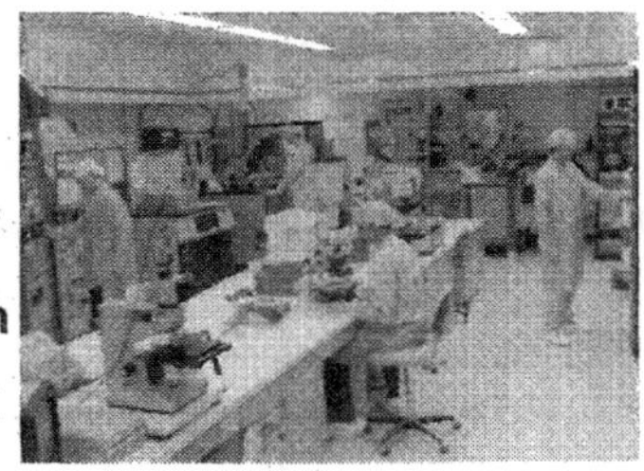

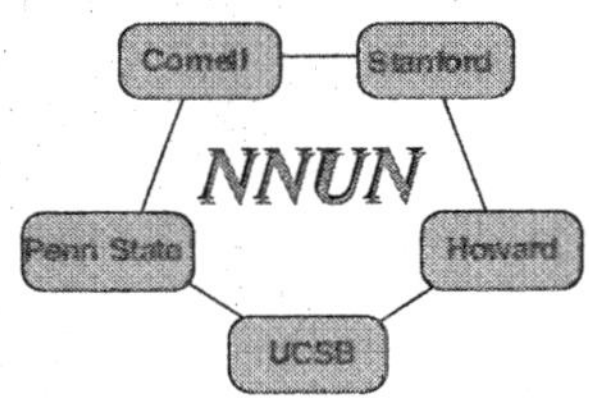

Figure 11.4. NNUN: What does it provide?

NNUN Nodes

- **Currently 5 nodes which provide basic micro & nanofabrication capability + particular expertise due to equipment and local users.**

CNF	• **General micro & nanofabrication** • **Electron beam lithography**
SNF	• **General micro & nanofabrication** • **Si devices and technology**
Howard	• **Wide band gap semiconductors**
PSU	• **General micro & nanofabrication** • **Nanofabrication in novel materials**
UCSB	• **Dry etching and III-V semiconductor structures**

Figure 11.5. NNUN nodes.

11.7.5 The Center for Quantized Electronics Structures (QUEST)

Contact person: E. Hu, University of California, Santa Barbara

QUEST, the Center for Quantized Electronic Structures (http://www.quest.ucsb.edu), is a National Science Foundation Science and Technology Center (www.nsf.gov/od/oia/stc), established in 1989 at the University of California at Santa Barbara (UCSB). QUEST's focus is a frontier field in nanostructure science and technology: the formation and study of "quantum structures." These are structures that generally have sizes sufficiently small that novel electronic, optical, and magnetic behavior emerges, which in turn can provide the basis for entirely new device technologies. QUEST integrates the research efforts of a multidisciplinary faculty from the departments of chemistry, chemical engineering, electrical and computer engineering, physics, and materials (see Figure 11.6). The work of QUEST spans the full range of growth and synthesis of quantum structures, characterization of their basic properties, and utilization of quantum structures in novel device schemes.

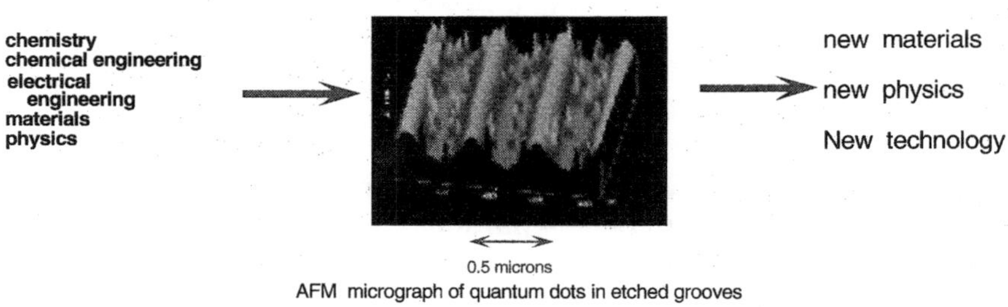

Figure 11.6. Science and technology at the atomic level.

QUEST Research

QUEST's focus has been the exploration of the novel physical and chemical properties of low-dimensional structures: where one or more of the structure's critical dimensions is below about 100 nanometers. It is at those lengths that the quantum mechanical nature of the material becomes more evident, and where the structure dimensions become equal to or less than important physical parameters such as the elastic mean free path for electrons. By controlling the critical dimensions of a structure, QUEST researchers hope to alter limitations to electronic transport and optical efficiency, in essence providing new materials that will sustain new device technologies. Although the majority of QUEST research has focused on quantum structures fabricated using compound semiconductors, the scope of the research is broadening to encompass a variety of other materials including oxides, superconductors and magnetic materials.

QUEST research strives to address the full range of issues necessary in spanning the science and technology of quantum structures. These include the following:

- The problems involved in the growth and fabrication of quantum structures
- The underlying physics and chemistry of these structures—what can be learned about their electronic, magnetic and optical properties
- The possible technological applications of quantum structures that may result once their behavior is well understood and controlled

The philosophy underlying QUEST's research strategy is illustrated in Figure 11.7, which represents the continuous, closely coupled interactions between fabrication, characterization, and simulation of quantum structures that takes place. The boxed text describes the critical challenges to be met in each area of research.

QUEST research is supported by world-class laboratories that include unique crystal growth and materials synthesis capabilities, a 3,500 sq. foot clean room with a Class-100 lithographic capability, and state-of-the-art fabrication processes. These latter include e-beam lithography and various dry etch and deposition processes. In addition, QUEST researchers utilize laboratories for low-temperature, optical, high-speed, and magnetic measurements, and they also have access to sophisticated surface science labs and make use of the Free Electron Laser at UCSB.

QUEST currently receives ~$3 million/year from NSF, along with additional industry and University funds for related work.

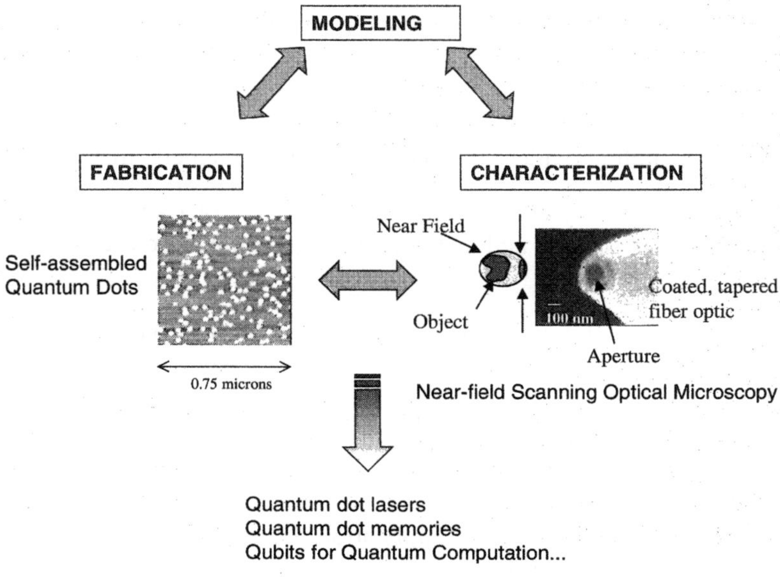

Figure 11.7. A continuous cycle of interactions.

11.7.6 Distributed Center for Advanced Electronics Simulations (DesCArtES)

Contact person: K. Hess, University of Illinois

The NSF-supported Distributed Center for Advanced Electronics Simulations (DesCArtES) consists of teams at Arizona State University, Purdue University, Stanford University, and the University of Illinois at Urbana-Champaign. Its mission is to attack key research and educational challenges for electronic devices and materials by complementing theory and experiment with large-scale computation. The focus, engineering-oriented but long-term, is on collaborative theme projects addressing (1) atomic scale effects in electronics, (2) silicon technology beyond the roadmap, and (3) optoelectronics. In addition to its core research efforts, DesCArtES provides outreach and leadership to the electronics research community through intellectual networking, network-based simulation and collaboration, and educational programs with input from industrial and Federal laboratories.

DesCArtES is co-directed by Karl Hess of the University of Illinois and Robert Dutton of Stanford. Umberto Ravaioli (Illinois) oversees research liaison and outreach activities, and Mark Lundstrom (Purdue) oversees educational outreach. To complement its core activities, partnerships have been formed with industrial researchers at a number of companies, including Lucent Bell Laboratories, Hewlett-Packard, Motorola, and Raytheon. An industrial advisory board has been formed to guide the center. DesCArtES has strong ties to several centers and organizations, including the National Computational Science Alliance (NCSA) for high-performance computing and the National Nanofabrication User's Network (NNUN) to connect with academic experimentalists. Collaborative projects are also underway with the Jet Propulsion Laboratory and the NASA Ames Research Center. A computational electronics "hub" that makes advanced simulation tools available to experimentalists and students has been deployed. The network already serves a worldwide user base. Figure 11.8 summarizes the core partnerships within DesCArtES.

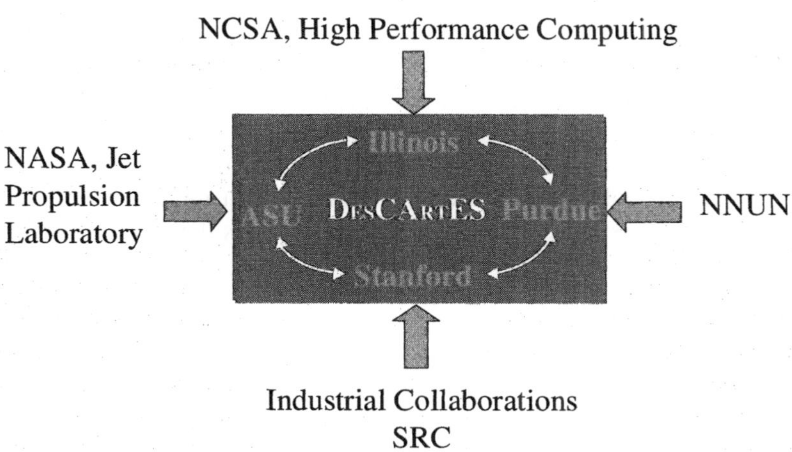

Figure 11.8. Distributed Center for Advanced Electronics Simulations (DesCArtES).

11.7.7 Nanoscience and Engineering at Materials Research Science and Engineering Centers (MRSEC Network)

Contact person: T. Weber, National Science Foundation

In 1999, nanoscale science and engineering is an area of focus in all 28 NSF Materials Research Science and Engineering Centers (MRSECs) funded by NSF. The centers are highly interdisciplinary, with the over 600 faculty participants coming from over a dozen academic departments. Approximately 75% of the annual budget of $45.5 million is targeted toward nanoscience and nanotechnology-related areas. The study of biomaterials, including biomimetic materials, which are synthesized based on examples provided by nature, is a very rapidly growing area of nanoscience and MRSEC research. Currently, extensive research in this area is carried out at eight centers, two of which are described below.

The work on abalone shell at the MRSEC at the University of California Santa Barbara draws faculty from materials chemistry, chemical engineering, physics, mechanical engineering, and molecular genetics fields. This group has been investigating the reasons that the abalone seashell is 3,000 times more fracture resistant than the basic calcium carbonate material that is the dominant shell ingredient. The group discovered that the secret lies in the polymer "glue" holding the layers of the shell together. Using atomic force microscopy techniques of pulling on single molecules of the polymer it has been shown that the polymer is made up of "knots" that unravel, one at a time, as an increasing force is applied. The stress is therefore relieved internally in the polymer molecule before the entire shell breaks apart. This discovery has allowed the UCSB researchers to propose the basic ingredient for making a "molecular" adhesive, with properties that mimic those of the natural product. This discovery is likely to have far-reaching impact in the technology of adhesives.

The University of Wisconsin, Madison, MRSEC is focused on nanoscale properties of semiconductors and high temperature superconductors. Through these efforts, which involve extensive materials growth and characterization capabilities, the center has also instituted an aggressive new program, funded out of MRSEC "seed" funds for innovative projects, on the fabrication of nanostructured surfaces as templates to study the growth of biological cells, in particular corneal epithelial cells. The focus and hypotheses for this "seed" project are noted in Figure 11.9. Although the project is still in its early stages, its researchers have been able to create synthetic templates that can serve as substitutes for the equivalent in a living organism. They have observed changes in the growth of cells based on the nanoscale substrate structure that are of fundamental importance in biodiversity and bioengineering.

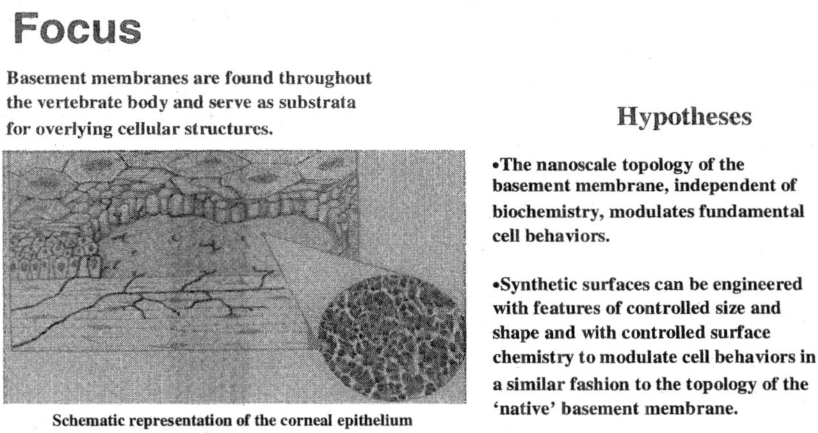

Figure 11.9. The influence of substrate topography on cell growth (©courtesy C.J. Murphy, Univ. Wisc.).

11.7.8 Nanotechnology at Sandia National Laboratories

Contact person: S.T. Picraux, Sandia National Laboratories

Sandia National Laboratories is a multiprogram Department of Energy (DOE) laboratory with 7,500 employees. Its principal mission is nuclear weapons stewardship. Sandia science and technology research supports a wide range of activities, including national security, nonproliferation, energy, and environmental programs. In accomplishing these tasks, the staff members interact extensively with industrial, academic, and government partners. Advances in nanoscience and technology are benefiting DOE research and development activities. The manipulation of the nanostructure of materials enables unique properties to be achieved for applications ranging from microlocks for weapons to high-efficiency photovoltaics, and from microchemical sensor systems to radiation-hardened microelectronics.

Integrated microsystems provide a striking example of the growing importance of nanotechnology. Microsystems are collections of small, smart devices that not only think (i.e., process information) but may also sense, act, and communicate. They combine microelectronic, photonic, micromechanical, and microchemical devices to create new generations of low cost miniature and highly reliable systems. Although they are built at the micron to centimeter dimensional scales, their performance depends on the control of materials properties at the nanoscale. Sandia's leadership in this emerging field is built on the ability to integrate this broad range of technologies. Across the wide range of activities from research to application, about 500 people are working in this area. Microsystems provide an excellent opportunity to combine nanotechnology advances with Sandia's inherent strengths in microfabrication.

To accomplish these tasks, special facilities and scientific expertise are maintained in the areas of nanoscience, microfabrication, and integration, including materials synthesis and processing; micro- to nanoscale probes; microelectronics; photonics; microsensors; microelectromechanical devices (MEMS); and computer, information, and systems science.

Examples of incorporating nanoscience into new technical capabilities for defense and energy applications at Sandia are wide ranging. For example, self-aligned monolayers provide dramatic improvements in surface tribology to reduce sticking and wear for MEMS devices. Vertical cavity surface emitting lasers (VCSELs) developed at Sandia use layered quantum well structures to produce highly efficient light sources for low power applications. Nanoclusters such as 3 nm diameter crystals of MoS_2 are being explored for their ability to photocatalyze the oxidation (destruction) of organic pollutants using only visible room light. Organically functionalized mesoporous structures are being integrated into micromachined devices on a centimeter-sized chip to provide thousand-fold chemical preconcentrators for on-chip analysis of chemical warfare agents.

The strength of such integrated capabilities is illustrated by the μChemLab™ project. In this exploratory project, a hand-held chemical sensing microsystem is being developed for detection of chemical and biological materials such as explosives and biological warfare agents. This system-on-a-chip approach involves preconcentration using nanostructured materials, separations with approximately one meter of spiral separating column embedded into a chip only 1 cm on a side, and then detection based on integrated optical fluorescence and/or piezoelectric acoustic wave detectors. The μChemLab™ depends upon integrating chemical, electronic, micromechanical, and photonic devices into microsystems. To achieve this broad integration goal, approximately 40 technical staff members from across the laboratory and an overall budget of $20 million (over 3 years) are currently focused on this project.

Increasingly we see that technical success depends on the ability of multidisciplinary teams to combine both the science and technology that cut across conventional disciplinary lines. Future technological advances in microsystems will depend upon discoveries in nanoscience and technology, combined with the ability to integrate technologies through low-cost, high-volume microfabrication methodologies (Figure 11.10), as can be done readily at a multipurpose national laboratory such as Sandia. Sandia's capabilities include, for example, 0.5 micron radiation-resistant Si IC design and fabrication, Sandia-developed five-layer MEMS device fabrication, growth and processing of photonic devices such as VCSELs, processing of novel new microchemical sensing devices, and a wide range of materials diagnostics, ranging from Sandia-developed new scanning probe techniques at the nanoscale to coupling first principles atomic scale modeling with microscale materials performance.

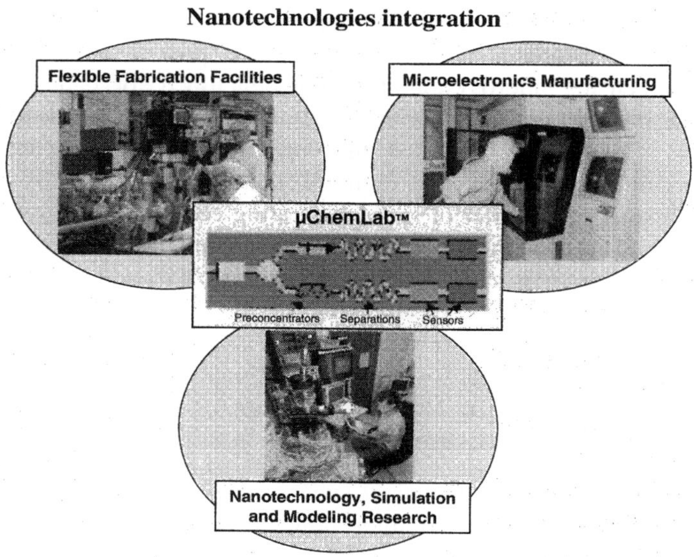

Figure 11.10. Nanotechnologies draw upon extensive multidisciplinary capabilities and a broad facilities base. The figure illustrates three key capabilities that are integrated together at Sandia: microelectronics manufacturing, flexible fabrication facilities, and nanoscale materials, simulation, and modeling research.

11.7.9 University of Notre Dame Center for Nanoscience and Technology

Contact person: J.L. Merz, University of Notre Dame

The Center for Nanoscience and Technology at the University of Notre Dame actively explores multidisciplinary fundamental concepts in nanoscience and engineering, with strategic emphasis on applications to unique functional capabilities. The center, established in 1998, integrates six research thrusts in molecular-based nanostructures: semiconductor-based nanostructures, device concepts and modeling, nanofabrication characterization, image and information processing, and function systems design. This effort cuts across four departments at Notre Dame: electrical engineering, computer science and engineering, chemistry and biochemistry, and physics, and it teams 24 senior faculty members along with their graduate students and postdoctoral researchers.

A major emphasis of the center is the concept of computing with quantum dots—quantum-dot cellular automata (QCA)—which is based on encoding binary information through the charge configuration of quantum-dot cells. The QCA notion has spurred further studies into nano-based cellular architectures for information processing that includes hierarchical functional design. The center also supports other initiatives in nanoscience and electronics, such as resonant-tunneling devices and circuits; photonic integrated circuits; quantum transport and hot carrier effects in nanodevices; and optical and high-speed nano-based materials, devices, and circuits.

The center has excellent on-site research facilities and capabilities including nanolithography and scanning tunneling microscopy; nanodevice and circuit fabrication; nanooptical characterization, including femtosecond optics and near-field scanning optical microscopy; electrical characterization at helium temperatures and in ten tesla magnetic fields; fifty gigahertz high-speed circuit analysis; and device and circuit simulation and modeling. In recent years, Federal grants received to support research in nanoscience and technology have totaled approximately ten million dollars, including two major grants from DARPA for Ultra molecular electronics ("Moletronics") programs, and several other awards from NSF, ONR, and the Army Research Office (ARO).

11.7.10 Nanophase Technologies Corporation: A Small Business Focused on Nanotechnology

Contact person: R.W. Siegel, Rensselaer Polytechnic Institute

Nanophase Technologies Corporation (NTC) was founded in late November 1989 by a scientist (R.W. Siegel) at Argonne National Laboratory (ANL) and ARCH Development Corporation (the technology transfer arm of ANL), and the University of Chicago, the DOE contractor for ANL. The company was a spin-off from a pioneering fundamental nanophase materials research effort in the Materials Science Division at ANL funded by the Department of Energy Basic Energy Sciences program.

Initial funding for NTC was supplied by ARCH, through its associated venture capital fund, and by the State of Illinois, through grants for new job creation. Subsequent funding was raised from a consortium of venture capital funds, and later also from high net worth private individuals and groups. The company went public with a successful IPO in late November 1997. An additional source of funding that was very important to NTC's development was an ATP grant from the Department of Commerce, which enabled the company to develop its patented physical vapor synthesis (PVS) process for manufacturing nanocrystalline materials in commercial quantities. This process was based on the laboratory-scale technology used at ANL from 1985 onward. NTC has also developed complementary nanoparticle coating and dispersion technologies, including its proprietary discrete particle encapsulation (DPE) process, as well as capabilities for superplastic forming of ceramic parts. Together, these technologies have enabled NTC over the past decade to enter a number of viable commercial markets. The company presently employs about 40 full-time workers (about 15 of whom hold advanced degrees) in its suburban Chicago facility.

NTC currently targets several markets: electronics (including advanced electronics, electromagnetic radiation protection, and advanced abrasives for chemical mechanical polishing); ceramic parts; specialty coatings and catalysts; and other technologically similar applications. In each of these market areas, NTC establishes collaborative relations with major corporate customers to develop and jointly implement nanoscale solutions for the customer's needs. In many cases, products developed to satisfy a particular vertical market need also have significant applicability across similar or horizontal markets. For instance, materials used in conductive coatings also have applicability for antistatic coatings and conductive strip carriers for color toners.

The applications for materials developed by NTC technology range from transparent protective coatings for CRT displays to highly engineered materials for chemical process catalysts. The NTC Web site provides current updates: http://www.nanophase.com. NTC is now focusing on, and will continue to emphasize, those applications where its nanoscale materials represent a technology breakthrough. As a nanomaterials company, NTC's continuing interest is to gather core technologies that provide the capability to service multiple major markets ranging from electronics to chemical processing.

11.7.11 Nanotechnology Infrastructure Capabilities and Needs in the Electronics Industry

Contact person: R.K. Cavin, Semiconductor Research Corporation

The electronics industry has a substantial interest in and history of exploring and exploiting nanotechnology in the development and fabrication of microelectronics. Metal lines as narrow as 10 nm can be printed in research, dielectrics as thin as 1 nm are being fabricated, and a variety of electron beam and atomic force metrologies are being used to characterize materials and structures at the atomic scale.

Future microelectronics opportunities for nanotechnology include engineering new materials integrated into conventional silicon chips (e.g., high permittivity, high-frequency permeability, high thermal conductivity, and high electrical resistivity). Other nanomaterials opportunities include self-assembly techniques related to new materials and nanometer-scale line printing. Opportunities also exist for nanotechnology to overcome scaling limits of current CMOS gate structures through the invention and development of radically new information processing technologies (e.g., FET replacement, ultra-high density memories, millimeter-wave devices). Applications that extend the capabilities of electronic systems can benefit from the innovative use of nanotechnology as well. These include mixed-function integration of new detectors, sensors, and optical and mechanical switches. Furthermore, hybrid integration of these devices (including digital and analog functions) also can be enabled by nanotechnology.

The electronics industry has an impressive array of capabilities and facilities related to nanotechnology. These include facilities for fabricating and characterizing complex nanostructures in electronic materials. Fabrication processes include epitaxial growth techniques (e.g., MBE, MOMBE, CBE, OMVPE, ALE), lithographic fabrication techniques (e.g., e-beam, EUV, X-ray) and characterization techniques (SEM, TEM, STEM, AFM, XPS, UPS, AES, EELS, etc.). The industry is also developing a new array of modeling tools to comprehend the quantum phenomena made possible by nanotechnology. An example of nanotechnology fabrication, the operation of quantum dot flash memory, is shown in Figure 11.11.

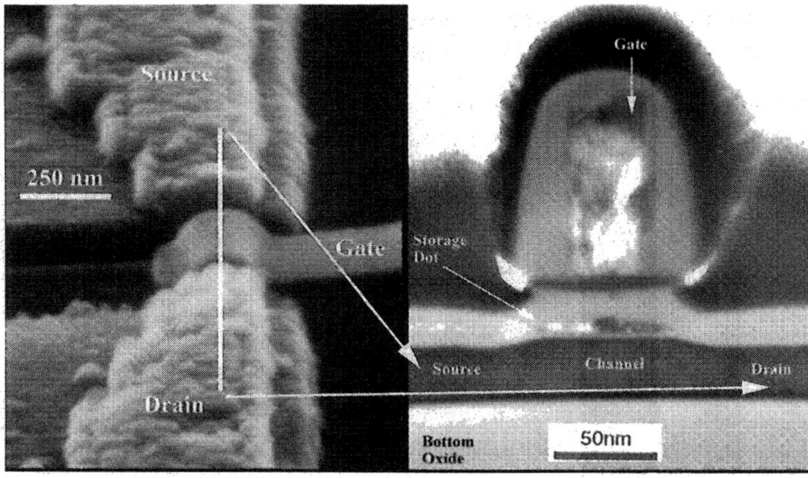

Figure 11.11. Room-temperature operation of a quantum-dot flash memory (reprinted with permission from Welser et al. 1997, ©1997 IEEE, courtesy of IBM).

11.7.12 Nanoscience and Nanotechnology at Lawrence Berkeley National Laboratory (LBNL)

Contact persons: M. Holm and M. Alper, Lawrence Berkeley National Laboratory

Lawrence Berkeley National Laboratory (LBNL) is a multiprogram DOE laboratory that conducts research addressing national needs in fields including the materials, chemical, earth, life, and environmental sciences and also energy efficiency, high energy, and nuclear physics. It is located adjacent to the University of California, Berkeley campus; 250 members of the Berkeley faculty lead Berkeley Lab research groups, 360 of their graduate students are trained and do their thesis research at LBNL, and 320 undergraduates are introduced to research in its laboratories. Each year, more than 2,000 guests come to work with Berkeley Lab staff and its unique research facilities. These include a wide variety of state-of-the art instruments developed at the laboratory and also five national user facilities, two of which have been designed, constructed, and operated by the DOE Office of Basic Energy Sciences: the Advanced Light Source, the world's brightest source of soft X-rays, and the National Center for Electron

Microscopy, home to the country's highest-resolution transmission electron microscope.

Berkeley Lab was one of the first to develop strong nanoscience programs and has a number of ongoing activities (Figure 11.12). The lab has been notably successful in linking chemists, physicists, biologists, and materials scientists in its efforts. In 1991, a program was begun to develop techniques for synthesizing nanocrystals of semiconductors and metals of controlled size. A theory component studies electronic and optical properties of nanocrystals and nanometer-size conducting polymers.

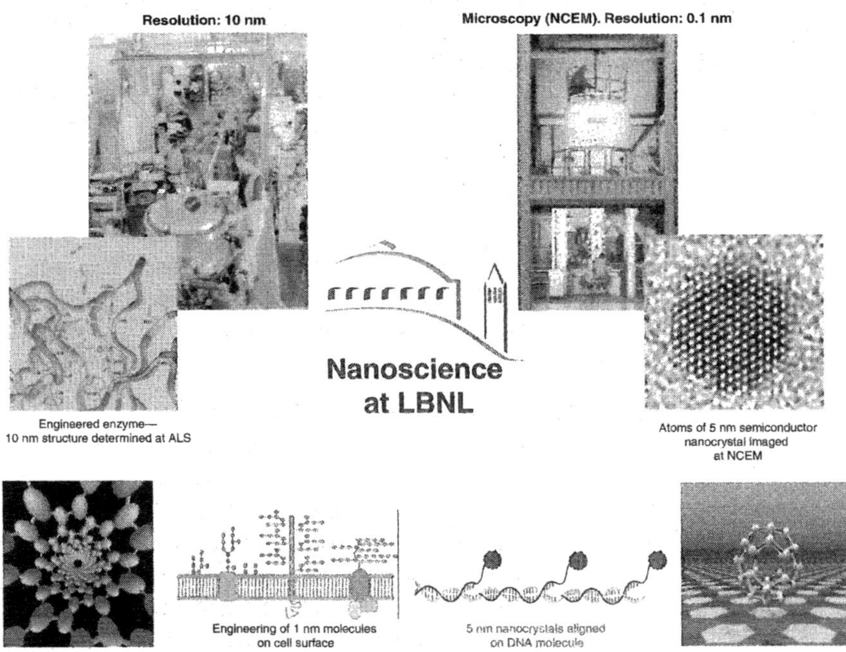

Figure 11.12. Nanoscience at Lawrence Berkeley National Laboratory.

Arrays of nanocrystals of defined spatial geometry have been fabricated at Berkeley Lab by attaching the crystals to strands of DNA of defined base sequence. A variety of techniques are employed to fabricate and study nanometer-size artificial magnetic structures. A team of experimentalists and theorists has predicted and synthesized nanostructures of carbon. They discovered the small C36 "buckyball" and nanotubes of rolled graphite that, depending on their structure, have either semiconducting or metallic properties. When linked, these tubes are postulated to behave as "nanodiodes." Lithographic techniques are used to build nanometer-size devices into which nanocrystals or nanotubes can be inserted. These are used to study the electronic and transport properties of these materials.

Investigators at Berkeley Lab have, for almost two decades, used optical methods to explore the dimensionality dependence of fundamental physical processes. Other groups are studying approaches for solving the difficult problem of introducing dopants into nanocrystals. A group is exploring the use of nanometer-size magnetic particles in high-sensitivity sensors.

Berkeley Lab was one of the first to develop a focused biomolecular materials program, recognizing that biological structures are nature's nanostructures. Enzymes are engineered through the introduction of specific alterations at the atomic level. Carbohydrates and related biomolecules are designed to control surface properties. Biologically inspired, nanometer-size polymers of specific structure and function, (dendrimers) are made "shape-persistent" with defined shape, chemical reactivity, and flexibility. They are being explored for their unique catalytic activity, light harvesting, energy transfer activity and other functions.

Tools are at the heart of cutting-edge research and the Berkeley program has a wide repertoire of advanced instrumentation. The Advanced Light Source, a unique third generation source of UV and soft X-ray photons, is utilized to give detailed structural and electronic information on, for example, polymers, semiconductors, layered materials, and proteins. The National Center for Electron Microscopy, with some of the most powerful microscopes in the world, provides images, for example, showing the uniform atomic distribution, faceting, and coatings of core structures of nanocrystals. In addition, LBNL's vast experience in developing and using scanning probe technologies (e.g., STM, AFM) and spectroscopies is of great value to the research groups. Instruments have been developed to operate at low temperature, at high pressure, with femtosecond time resolution, and with single molecule resolution.

11.7.13 The Social Impact of Nanotechnology: A Vision to the Future

Contact person: J. Canton, Institute For Global Futures

To state that nanotechnology will have a profound impact on society would be a gross understatement. If we had tried to explain to an eighteenth century person how a television or a computer worked, let alone the Internet, we would have been considered mad. These innovations, though they seem routine today, offer only a hint of what is to come. We lose track of the fact that there have been more innovations that have changed society in the past fifty years than in the previous five thousand years. New tools such as computers and networks have empowered fantastic innovation. But all that has been accomplished in the past will seem small in the face of the changes brought by nanotechnology in the future.

Nanotechnology is a comprehensive design science that will give us powerful new tools that may change every aspect of society. This technology will place in our hands the ability to design matter at the molecular and atomic level. We will be able to eventually fabricate existing products on-demand and more inexpensively. More interestingly, using natural principles and processes, we will design products that have never existed in nature.

Two critical factors will emerge in society: how fast people adapt and how smart they become about the application of nanotechnology solutions. These factors will determine the competitiveness of individuals, organizations, and nations. Nanotechnology knowledge, how to develop it, and how to use it, will become a strategic asset. Those societies that support nanotechnology education, research, and development the fastest will thrive in the new millennium. Just as today's digital technology drives personal and business success, nanotechnology will dominate the twenty-first century.

The general societal impact of nanotechnology will be felt in some of the following ways:

- Consumer and industrial products will be smaller, more durable, smarter, faster and less expensive due to the super-efficiency of nanoengineered materials and manufacturing.

- Healthcare will become less expensive, more accessible, and more effective at preventing disease, replacing human parts, and enhancing life. New drugs and diagnostic devices will be available. Nanobiology will empower people to live longer and healthier lives. Synthetic tissue and organs, genetic and biomolecular engineering, and "directed evolution" will emerge.
- Embedded intelligence will be everywhere: from chips in paper, to clothing that talks, to cars that self-generate their own energy, to Internet-ready devices that combine the functions of a TV, telephone, and computer. Everyone, anyplace, anytime, will be interconnected.
- Business will need to retrain workers with the skills necessary to survive in a new economic reality based on nano-products and nanotechnology knowledge. This is a paradigm shift that will demand knowledge of nano-engineering. Just as the Internet is forcing every business to become an e-business, every business in the twenty-first century will become a nano-business.
- Education will need to change entirely to address the fast development of nano-industries. The coming generations will need to be trained in nanoscience. Every educational discipline from engineering to chemistry to physics will require new learning.
- Nano-energy will make for both a cleaner and more fuel-efficient world. New engines for vehicles, fuel cells, and transportation will be possible.
- Food production will become nano-engineered, providing a bounty of inexpensive, nutritious, and appealing culinary choices that are less dependent on nature than on nanoscience.
- Work and careers will be deeply affected as people retool for a nanotechnology-enhanced economy that is less product-driven and more service and knowledge driven. Lifestyle choices will become more varied as nanotechnology changes the global economics of supply and demand.
- New choices will be developed for the augmentation of cognitive processes, and increase of physical and sensory performance.
- The virtual asset-based economics of living in a society that is dominated by nanotechnology will quickly reward those

individuals and organizations that hold the intellectual properties to this new technology.

Nanotechnology's impact on society will be comprehensive, touching all aspects of lifestyle, quality of life, and community. Inevitably, nanotechnology will give people more time, more value for less cost, and provide for a higher quality of existence. The convergence of nanotechnology with the other three power tools of the twenty-first century—computers, networks, and biotechnology—will provide powerful new choices never experienced in any society at any time in the history of humankind.

The social impact of nanotechnology will need to be a managed-change process. Never has such a comprehensive technology promised to change so much so fast. A national policy on nanotechnology should include responsible oversight. Those organizations and citizens who are unaware of this impeding power shift must be informed and enabled so that they may adequately adapt.

11.8 REFERENCES

Aksay, I.A., et al., eds, 1992. Hierarchically structured materials. *Materials Research Society Proceedings* #255.

Csurgay, A.I. 1997. The circuit paradigm in nanoelectronics. In *Proceedings of the ECCTD'97*, 240-246. Budapest.

Drexler, E.K., C.H. Peterson, and G. Pergamit. 1993. *Unbounding the future: The nanotechnology revolution.* New York: Quill Books.

FED (Future Electron Devices program, Japan). N.d. Web page: http://www.iijnet.or.jp/fed-www/.

Freitas, R.A. 1999 (in press). *Nanomedicine.*

Gleiter, H., 1989, Nanocrystalline materials. *Progress in Materials Science* 33: 223-315.

MEL-ARI (Microelectronics Advanced Research Initiative of ESPRIT). N.d. Web page http://www.cordis.lu/ esprit/src/melari.htm.

Merkle, R.P. 1994. Self-reproducing systems and low cost manufacturing. In *The ultimate limits of fabrication and measurement.* Ed. M.E. Welland et al. Dordrecht: Kluwer.

Porod, W. 1997. Quantum-dot devices and quantum-dot cellular automata. *Int. J. of Bifurcation and Chaos* 7.

Porod, W., ed. 1998. Special issue on computational electronics. Papers presented at the Fifth International Workshop on Computational Electronics, Notre Dame, *VLSI Design* 8(1-4).

Crandall, B.C., ed. 1996. *Nanotechnology: Molecular speculations on global abundance*. Cambridge, Mass.: MIT Press.

Jortner, J. and M. Ratner, eds. 1997. *Molecular electronics: A 'chemistry for the 21st century' monograph.* Oxford, U.K.: Blackwell Science Ltd.

Regis, E., and M. Chimsky. 1996. *Nano: The emerging science of nanotechnology.* Little Brown and Co.

Siegel, R.W., E. Hu, and M.C. Roco, eds. 1999. NSTC (National Science and Technology Council) Report. *Nanostructure science and technology.* Baltimore: International Technology Research Institute, World Technology (WTEC) Division. Web site: http://itri.loyola.edu/nano/IWGN.Worldwide.Study/. Also published by Kluwer Academic Publishers (1999).

Smalley, R.E. 1995. Nanotechnology and the next 50 years. Presentation to the University of Dallas Board of Councilors, December 7, 1995. http://cnst.rice.edu/dallas12-96.html.

Von Neumann, J., and A.W. Burks. 1966. *Theory of self-reproducing automata.* University of Illinois Press.

Welser, J.J., S. Tiwari, S. Rishton, K.Y. Lee, and Y. Lee. 1997. Room temperature operation of a quantum-dot flash memory. *IEEE Electron Device Lett.* 18: 278.

Whitesides, G.M., J.P. Mathias, and C.T. Seto, 1991. Molecular self-assembly and nano-chemistry: A chemical strategy for the synthesis of nanostructures. *Science* 254: 1312-1319.

Chapter 12

AGENCY FUNDING STRATEGIES

Contact persons: R.S. Williams, Hewlett-Packard; G.D. Stucky, UCSB

12.1 VISION

There is no question that nanotechnology in its broadest sense will be a dominant force in our society in the early decades of the next century. The primary questions are how soon this destiny will arrive, what benefits and risks it presents, and how the United States can be in the best position to guide and capitalize on the development of nanotechnology for the benefit of Americans and for the world. U.S. funding agencies have both the opportunity and the obligation to seed the scientific efforts that will nurture nanotechnology to the point where they can realize their beneficial intellectual, economic, and societal potentials. The challenge for the funding agencies is to formulate a long-term and sustainable strategy that promotes the healthy development of nanotechnology within the constraints imposed by the annual cycle of the Federal budget. It is strongly recommended that this process begin with the creation of a high profile National Nanotechnology Initiative, which will have the short-term goal of doubling the Federal Government's present investment in nanotechnology research in fiscal year 2001.

12.2 CURRENT FUNDING PRACTICES

Nanotechnology research in the United States has developed thus far in open competition with other research topics within various disciplines. This is one reason that U.S. nanotechnology research efforts tend to be fragmented and overlap among disciplines, areas of relevance, and sources of funding (Roco 1999). This situation has advantages in establishing competitive paths in the emerging nanotechnology field and in promoting innovative ideas, and some disadvantages for developing systems applications. As the far-reaching consequences of nanotechnology R&D has begun to be appreciated within the scientific community and various Government agencies, interest has grown in focusing national resources on stimulating cooperation, avoiding unwanted duplication of efforts, and building a supporting infrastructure that will better position the United States to lead and benefit from the revolution that is coming. Twelve funding/research agencies established an informal group in 1997 in order to enhance communication and develop partnerships. The National Science and Technology Council (NSTC) formally established the Interagency Working Group on Nanoscience, Engineering, and Technology (IWGN) on September 23, 1998.

An inventory of current activities and R&D future needs was assembled at the workshop held by the IWGN on January 27-29, 1999. This inventory is presented herein. The agencies participating in the working group are the departments of Commerce, Defense, Energy, and Transportation (DOC, DOD, DOE, DOT), the National Aeronautics and Space Administration (NASA), the National Institutes of Health (NIH), and the National Science Foundation (NSF), with NSF, DOD and DOE making the largest investment in nanotechnology in fiscal year 1999. Other agencies with nanotechnology-related activities that may be added in the future include the Department of Justice (DOJ) (with interest in forensic research, high-performance computing and database management), the Environmental Protection Agency (EPA) (with interest in measurement and remediation of nanoparticles in air, water, and soil), the Treasury Department (with interest in special colloidal suspensions at the Bureau of Engraving and Printing).

Section 12.7.1 outlines the main R&D themes, current focused programs and initiatives, as well as research opportunities at the seven major Government funding departments and agencies. The estimated total funding from the U.S. Federal agencies in fiscal year 1999 is approximately $255 million (based on the IWGN survey in June 1999, see Section 12.7.2). The projected nanotechnology-related needs of all participating departments and agencies for fiscal year 2001 total roughly double the amount of the current budget. For each of the seven major funding departments and agencies, this section provides a concise summary of the agency's current major interests in nanotechnology, and the themes and modes of R&D support proposed for increased funding in fiscal year 2001.

Department of Commerce (DOC, including the National Institute of Science and Technology, NIST)

1. *Current major interests in nanotechnology:* Measurement science and standards, including methods, materials, and data; development and acceleration of enabling commercial technologies through industry-led joint ventures (Advanced Technology Program—ATP). The current budget for nanotechnology is divided between measurement and standards research, and ATP cost-shared awards to U.S. industry.
2. *Themes and modes of proposed R&D support for fiscal year 2001:* Develop the measurement and standards infrastructure to support U.S. industry development and commercialization of nanotechnology; and perform economic and foreign assessment studies. Major themes in fiscal year 2001 include nanodevices and biotechnology for quantum level measurement and calibration; magnetic measurements and standards research; nanoscale characterization—measurement systems, approaches, and algorithms; standard data and materials; and nanoscale manipulation for synthesis and fabrication.

Department of Defense (DOD)

1. *Current major interests in nanotechnology:* Information acquisition, processing, storage, and display; high performance, affordable materials; and bioengineering for chemical and biological warfare defense, casualty care, and human performance monitors.
2. *Themes and modes of proposed R&D support for fiscal year 2001:* Investigator projects; focused programs and initiatives (e.g., the Multidisciplinary University Research Institute—MURI program, instrumentation grants, Defense Advanced Research Projects Agency—DARPA programs); DOD service laboratory programs; and cooperative research and development agreements between laboratories and commercial ventures. Major themes and new programs include advanced processes and tools; nanoelectromechanical systems (NEMS), with focus at DARPA;

biocentric research, where nano is part of the Office of Naval Research (ONR) overall program; and MURI topics focused on nanotechnology.

Department of Energy (DOE)

1. *Current major interests in nanotechnology:* Basic energy science and engineering, including experiments, diagnostics, fabrication and modeling, energy efficiency, defense, environment, and nonproliferation. The largest expenditures in the current budget include materials, chemistry, defense-related projects, and engineering.
2. *Themes and modes of proposed R&D support for fiscal year 2001:* Capital development at national labs; secondary funding of universities for collaboration with DOE labs; programs to encourage national labs to work with other Government agencies and industry; and 2-3 laboratory user facilities. Increased funding is needed to support both a network of research user facilities at four national laboratories and academic research for energy- and environment-related topics.

Department of Transportation (DOT)

1. *Current major interests in nanotechnology:* Nanostructured coatings; sensors for physical transportation infrastructure; and smart materials. DOT incorporates the results of nanotechnology R&D into its more focused R&D programs, without having specialized departments for nanotechnology R&D.
2. *Themes and modes of proposed R&D support for fiscal year 2001:* Efficient incorporation of research results into more focused DOT research and technology activities: nanostructured coatings for metallic surfaces to achieve super-hardening, low friction, and enhanced corrosion protection; "tailored" high-performance materials with reduced life-cycle costs, greater strength-to-weight, and longer service life for vehicles and infrastructure; "smart" materials that monitor and assess their own status and health and that of systems and subsystems; monitoring and remediation of oil spills and other hazardous materials incidents; studies of the implication of advances in nanotechnology for the next-generation of transportation professionals.

National Aeronautics and Space Administration (NASA)

1. *Current major interests in nanotechnology:* Lighter and smaller spacecraft; biomedical sensors and medical devices; powerful, small, lower power consumption computers; radiation hard electronics; and thin film materials for solar sails.
2. *Themes and modes of proposed R&D support for fiscal year 2001:* Three laboratories—Jet Propulsion Lab (Pasadena), Ames Research Center, and Johnson Space Center (Houston)—and academic research on space exploration topics. Research needs have been identified in the following areas: techniques for manufacturing of single-walled carbon nanotubes for structural reinforcement, electronic, magnetic, lubricating, and optical devices, chemical sensors, and biosensors; tools to develop autonomous devices that articulate, sense, communicate, and function as a network, extending human presence beyond the normal senses; and robotics using nanoelectronics, biological sensors, and artificial neural systems.

National Institutes of Health (NIH)

1. *Current major interests in nanotechnology:* Biomaterials (e.g., materials-tissue interfaces, biocompatibility); devices (e.g., biosensors, research tools); therapeutics (e.g., drug and genetic material delivery); and infrastructure and training.
2. *Themes and modes of proposed R&D support for fiscal year 2001:* Fund academic research, small business research, and in-house studies on nanobiotechnology, including the following topics: advances in biomaterials; clinical diagnostic sensors; genomic sensors; and nanoparticles and nanospheres for drug and gene delivery.

National Science Foundation (NSF)

1. *Current major interests in nanotechnology:* Fundamental academic research on novel phenomena, synthesis, processing, and assembly at nanoscale; generation of new materials by design; biostructures and bio-inspired systems; system architecture at nanoscale; instrumentation and modeling tools; high-rate synthesis

of nanostructures and scale-up approaches; infrastructure and education; university-industry collaborations.

2. *The research themes and modes of proposed R&D support for fiscal year 2001:* The main research themes are: (a) nano-biotechnology, including biosystems, bio-mimetics and composites; (b) synthesis and processing of nanostructures "by design," and investigation of new phenomena and processes at nanoscale; (c) integration of nanostructures and nanodevices into systems and architectures, including multiscale and multiphenomenal modeling and simulation; and (d) investigation of environmental processes at nanoscale and at long time scales, including studies of the interactions between biological, organic and inorganic structures. Increase funding is envisioned for individual academic research and for centers/networks awards (ERC, MRSEC, Science and Technology Centers, and the National Nanofabrication Users Network).

12.3 GOALS FOR THE NEXT 5-10 YEARS: BARRIERS AND SOLUTIONS

Principal Goals and Challenges

The cardinal goal defined by members of the IWGN is that U.S. Government funding agencies must foster an enduring nanoscale science and technology culture that can in turn nurture industrial enterprises on the 10-20 year timeframe.

The major barrier to this goal is to convince decision makers that nanotechnology is important enough to warrant the special attention required to provide and maintain a sufficient funding base over the long term for this emerging and rapidly growing set of disciplines. Informing and educating decision makers is primarily the responsibility of the scientific, technical, and business communities, since the funding agencies themselves are prohibited from activities that may appear to be lobbying. The January 1999 IWGN "Vision for Nanotechnology" workshop and this report, as well as many previous workshops and their reports, are parts of the educational process.

In addition, a nanotechnology agenda will have to be supported by a broad coalition of scientists, engineers, and others in order for the cardinal goal to be achieved. The message of the IWGN has to be clear: Exploring the promise and exploiting the potential of nanotechnology is a long-term investment that will bring enormous societal and economic benefits, primarily to those most able to innovate and capitalize on the opportunities as they arise. To be either the leader or a fast follower in nanotechnology, the United States will have to be strongly engaged in the effort. Thus, the immediate goal is to establish a national nanotechnology initiative.

After achieving initial recognition of the importance of nanotechnology, the next major challenge will be to implement a sustainable long-term strategy. The Federal budget operates on an annual cycle, priorities change, and institutional memories are short. A significant danger to this endeavor is that the difficulty of the task will be underestimated during the early stages, and an inability to quickly produce the astounding advances so often hyped in the popular press may cause a backlash in the public and in Congress against long-term support for nanotechnology. Funding agencies must resist the temptation to rush into misguided development programs before the necessary science and technology base exists to identify realizable goals.

A further problem for nanotechnology development will be long-term competition for limited resources from the legitimate interests of other scientific and technology groups, which will argue persuasively for funding increases in their own areas. The proposed National Nanotechnology Initiative will provide a much needed short-term infusion of funding into the nanotechnology research community that will be effectively absorbed and utilized, ensuring that the best ideas are supported and drawing even more talent into the field. However, a long-term commitment with a steadily rising funding profile is necessary to establish a vigorous nanoscale science and technology community. This will require a funding strategy that will have to be reintroduced annually into the Federal budget process and continually supported by a broad range of the technical community.

Ancillary Goals and Challenges

The funding agencies will also need to encourage new modes of research and educational models to create a vigorous nanotechnology culture, as well as to provide the funding for an appropriate physical infrastructure and to maintain the research community for at least a decade. Eventually, the technology will evolve to the point where private industry becomes the dominant source of nanotechnology jobs, but there needs to be a supply of skilled workers ready for industry when industry is ready for commercialization.

Because of the shear breadth of nanotechnology, no single person or traditional discipline can encompass the range of skills and knowledge required for dramatic breakthroughs. Thus, small but agile teams of transdisciplinary researchers are likely to be best suited for innovation. However, this runs counter to the present structure for performing research, in which either an individual investigator or a large group of investigators (a center) from closely related fields are supported by disciplinary divisions within the various funding agencies. To encourage creativity, discovery, and invention in this mostly exploratory phase of nanoscience and engineering, a large proportion of the grants for nanotechnology research should be intended for small groups of one to four principal investigators, usually representing different disciplines (e.g., physics, biology and computer architecture) and/or institutions (including academe, national labs, and industry). This requires a commitment from the principal investigators to engage their colleagues and learn how to communicate across intellectual or organizational boundaries. Proposals from such groups should be reviewed by transdisciplinary panels that are instructed to take chances if the potential pay-off from a proposal is seen to be large. The primary metric for renewal of such proposals should be accomplishment.

There should also be a broader range of educational opportunities for students coming into nanotechnology areas. The students must gain in-depth knowledge in one subject, but they also need to develop breadth by being able to transcend geographical location, institution, and discipline. The problem with this goal is that most graduate students in technical areas are funded by the grants to their research advisors, and thus they are tied to a specific discipline and location because their mentors cannot afford to pay for students who are not in their labs. Thus, there should be a significant number of nanotechnology fellowships and training grants that will give the best students the ability to craft their own education by specializing in one area but having the opportunity to work with one or more other mentors. This will further encourage a practice that is already occurring, since much of the current transdisciplinary nanotechnology research efforts are actually initiated by students who realize the benefits of working with more than one advisor. Programs that encourage intermingling among science, engineering, and business disciplines should also be supported strongly, since grooming future technically competent entrepreneurs is at least as important as educating future professors and researchers. Nanotechnology workshops focused on graduate students with primary expertise in a large number of different disciplines should be held that would allow them to see and understand the bigger picture and encourage them to communicate across boundaries.

Funding agencies must also ensure that there is a sufficient physical infrastructure available for nanotechnology research to flourish. This requires a broad range of facilities, from the scanning probe microscopes in each investigator's lab to the expensive high-resolution electron and ion beam microscopes and lithography systems that require institutional support. One major impediment to this goal is the fact that funding agencies award grants for the purchase of major equipment items but do not always provide adequate resources for operation and maintenance. In general, funding agencies should consider a significant contribution toward running and repairing a major instrument to be part of the equipment grant to a university. Industry should also contribute to these costs, especially where it derives benefit from access to those instruments. The National Nanofabrication Users Network (NNUN) provides an example of a group of regional facilities based in host universities that provide access to equipment and expertise to outside users. However, the number of such facilities needs to increase significantly from the current number of five.

There must be enough funding now to support the best nanoscale science research in the current academic and national laboratory groups. However, there has to be a clear understanding that although industry will hire excellent technically trained people with almost any background, there will be very little work in industry on nanotechnology until it is near (probably within three years) to having a significant market impact. For many, if not most, areas within the nanoscale sciences, this will be at least ten years in the future. In the meantime, there must be a fertile ground where the best young researchers trained in nanoscale science can continue to contribute to and advance the field. This "major league" for nanotechnologists will be primarily in universities and national labs for the next decade. Thus, the funding profile for university grants and national labs in nanotechnology must increase at a rate that will encourage the best young researchers to stay in the field and allow them to build up their own research programs. This indicates that the funding profile for nanotechnology during the next decade must increase at a rate significantly higher than inflation, and that it may require at least ten years to reach a steady state. The first two products to come out of the early stages of Government funding will be trained people and scientific knowledge. There must be a critical mass of both before the development of a technology and intellectual property can occur. Once these become compelling, then actual products, manufacturing infrastructure, and high paying jobs will emerge that will repay the investments that have been made in this area.

12.4 SCIENTIFIC AND TECHNOLOGICAL INFRASTRUCTURE

By its very nature, nanotechnology is an extremely transdisciplinary area. The major portion of the research funding should be concentrated in small groups, essentially one to four principal investigators, dedicated to a particular project; however, it is also important to identify a few "grand challenges" that will require larger centers and consortia where ideas can bridge traditional intellectual, organizational, and geographical borders. There should not be a specific formula around which these centers and consortia are constructed, but some of them should involve collaborations among academic, national, and industrial laboratories. Industrial groups should be required to contribute some of the resources if they participate in these efforts, but the funding agencies should provide some support to research programs in industry (for instance by funding collaborative activities in partnering universities) to act as leverage to their commitments. In addition, flexible fellowships and training grants are necessary that will allow students and postdoctoral fellows a considerable amount of freedom to move between academic disciplines and geographic locations.

The issue of equipment infrastructure must be addressed. The most common instruments will probably be various types of scanning probe, electron, and ion microscopes. On the one hand, there has to be an understanding that a single research group can easily have a need for several different scanning probe microscopes, since there are now many different types, each optimized for a different task. On the other, the best electron and ion microscopes are very expensive and costly to maintain, and a means should be provided for universities to acquire, maintain, and operate such systems. There will also be a need for a wide range of facilities and more traditional instruments, ranging from synchrotron radiation and neutron sources, X-ray diffractometers, all types of spectrometers, and computational facilities to both handle the processing of massive amounts of data and carry out the crucial modeling and simulation work needed to advance the field rapidly. Since the emphasis for most of the groups performing nanotechnology research needs to be on the science and not the equipment, a larger number of shared laboratories and regional facilities should be funded and staffed. As proposed in Chapter 3 of this report, some of these facilities should be charged with the responsibility to develop new instrumentation, essentially to address the grand challenge of nondestructively determining the three-dimensional elemental and chemical state map of a system with sub-nanometer resolution and one atom per cubic nanometer sensitivity.

There should also be a few multi-technology engineering demonstration centers. These would be pilot fabrication facilities to try out new manufacturing ideas, such as methods to attain large-scale production of carbon nanotubes or semiconductor nanocrystals. An example of such a facility is the University of Barcelona-Xerox Laboratory for Magnetics Research, which has been initiated by the University of Barcelona, Xerox Corporation, and a number of smaller companies to minimize the time for transition from discovery to commercialization. Such a facility would be open to use by a variety of groups, from academic to large industry, on a fee basis according to the nature of the group involved.

The issue of information sharing is paramount: an agency and specific funding must be identified to foster communication of ideas and results among the various subfields within nanotechnology. One approach would be for an agency such as NIST to sponsor a nanotechnology-specific information clearinghouse and maintain an up-to-date database addressable via the Web, as well as sponsor workshops that involve younger researchers. Professional science and engineering societies should also take on the challenge of improving communication between the disciplines they traditionally represent and other contributors to nanotechnology, both nationally and internationally. The societies should provide transdisciplinary nanotechnology forums that enable creative exchange of ideas; commingle university, government and industrial researchers; provide tutorials that breakdown barriers between disciplines (e.g., the biological and physical sciences); and educate the public.

12.5 R&D INVESTMENT AND IMPLEMENTATION STRATEGIES

In any discussion of implementation strategies, there is a significant tension between what is the ideal path to follow and what is practical or realistic to attain. On the one hand, nanotechnology will require a long-term and coordinated commitment from the funding agencies, but on the other hand, the budgets for the funding agencies are set annually and balance competing interests. The responsibility for alerting budget planners to the opportunities and risks presented by the emergence of nanotechnology rests primarily with the scientific and technical community, but the funding agencies are best suited to provide the framework for increasing the priority for nanotechnology research. In order to have a strategy to implement, those working in the area of nanotechnology will first have to convince decision makers that this is truly a national priority.

To obtain the high-level attention necessary for a sustained investment, the fact that nanotechnology has reached a critical stage requiring immediate attention will be best articulated through the creation of the National Nanotechnology Initiative. A dramatic increase in funding for nanotechnology, in particular a doubling of the investment portfolio, from the current $255 million/year in 1999 to on the order of $500 million/year, is an important short-term goal. However, in order to satisfy the strategic vision for creating a viable nanotechnology culture from which new industries can emerge, that level of investment will have to double again over a period of five to seven years. The funding agencies will have to cooperate and coordinate their efforts, since the needs of the field will evolve with time.

The first priority is to ramp up funding for small groups in universities. The number of grants must increase, since only a small fraction of worthy proposals are currently being funded, but it is also important to increase the size of individual grants to make sure investigators are adequately supported. Next, programs in U.S. Government (national) and private laboratories should grow to provide the critical mass of researchers necessary to have a vibrant field and to create the broad technology base required for invention to emerge. Finally, new businesses and industries should be seeded through programs such as SBIR grants and access to regional centers and technology demonstration facilities.

Will a National Nanotechnology Initiative be a wise investment, and can it be effectively absorbed and utilized by the research community? Nanoscale science and engineering knowledge is exploding worldwide because of the availability of new investigative tools and interdisciplinary synergism, and is driven by emerging technologies and their applications. New experimental and modeling tools have opened additional windows of research opportunity. The publication rate in nanoscience is doubling every two to three years, and the number of revolutionary discoveries can be expected to accelerate over the next decade. As indicated in the body of this report, the discovery and development of new nanoscience principles will affect existing and emerging technologies in almost all industry sectors and application areas, including computing and communications, pharmaceuticals and chemicals, environmental technologies, energy conservation, manufacturing, and healthcare-related technologies. Because of the highly competitive nature of nanotechnology research and the potential for high economic and social return on investment, the need to establish a national initiative is compelling.

The reported Federal Government expenditure for nanotechnology in fiscal year 1997 was approximately $116 million (Siegel et al 1998; summarized in Siegel et al. 1999). Nanotechnology as defined there only included work to generate and use nanostructures and nanodevices; it did not include the simple observation and description of phenomena at the nanoscale that is part of nanoscience. Utilizing the broader definition, the Federal Government expenditure is estimated to be about $255 million for fiscal year 1999. However, a much greater investment would be utilized effectively, and in fact many promising opportunities are not being pursued now because of lack of resources. The funding success rate for the small-group interdisciplinary research program, NSF's fiscal year 1998 "Functional Nanostructures" initiative, was only about 13%, about 1/3 of the proposals of high quality that could have been funded (significantly lower if one considers the limitation of two proposals per university imposed in that initiative). The success rate for the DOD 1998 MURI initiative on nanostructures was 17% (and only 5% if one starts with the number of white papers submitted to guide proposal development). Since one-third of proposals submitted are usually considered to be "very good" to "excellent" and thus deserving of funding with priority, the number of high-quality ideas worth pursuing significantly exceeds the current investment level.

A rough estimate of the interest in nanotechnology can be made from the total number of scanning probes purchased, approximately 4,000 in the United States. One instrument can support two full-time research efforts (considering that experimental design, data analysis, and reporting take up at least half the time of a research effort) at a level of about $300,000/year for both personnel and equipment. (This probably somewhat overestimates the university laboratory cost and underestimates the government/industrial laboratory cost.) On this basis, without factoring in the contributions from theorists and researchers using other analytical equipment (such as high-resolution electron microscopes, near-field visualization instruments, and small angle neutron scattering), the capacity of the United States to perform nanotechnology R&D is greater than $1.2 billion/year. Doubling the current R&D expenditures in fiscal year 2001 will satisfy less than one-half of that research capacity, and thus ensures that many of the best research groups and best ideas are funded. Also, since obtaining a grant in nanotechnology will be more accessible, many more strong researchers will be drawn into and enrich the field.

12.6 PRIORITIES AND CONCLUSIONS

Nanotechnology is an emerging set of fields that will have enormous societal and economic consequences in the next century. These fields have reached a critical stage where their potential can be reasonably anticipated but has not yet been realized. In order to accelerate the development of nanotechnology to ensure that the United States is at or near the lead in as many of the fields a possible, a significant increase in the level of the effort and a reorganization of the current research culture are required. To achieve such major changes, a National Nanotechnology Initiative is proposed. A short-term doubling of the national investment in nanotechnology, followed by a longer-term redoubling will provide the resources required sustaining the high quality of the research currently being performed and implementing important new initiatives. The funding agencies will have the responsibility to encourage new collaborative research modes, provide broader educational opportunities, support a national infrastructure to provide open access to instrumentation, techniques, and information, and overall to sustain the nanotechnology R&D community during its developmental stage.

12.7 ATTACHMENTS: ILLUSTRATION OF FUNDING THEMES

12.7.1 General Themes and Initiatives on Nanotechnology at DOC, DOD, DOE, DOT, NASA, NIH, and NSF

Contact persons: M.C. Roco and J. Murday, for IWGN

Presented below is a summary of the main research and education themes, focused programs and initiatives, and R&D opportunities identified at the Federal departments and agencies with the largest contributions to nanotechnology.

1. DOC: Department of Commerce, especially the Role of the National Institute of Standards and Technology (NIST) in Nanotechnology

Main Themes

Nanotechnology has profound implications for NIST's mission to support the U.S. measurements and standards infrastructure. NIST's research interest in nanotechnology is due to its mandate to develop low-cost, widely available primary standards and atomically accurate measurement systems. Nanotechnology-based standards have the potential to greatly improve the efficiency of the U.S. standards system by reducing the currently laborious traceability chains needed for reliable products.

U.S. industry is also interested in nanotechnology, including the latest fundamental discoveries, as evidenced by the marketing of new nano-based products from computer disk storage to cosmetics and the many proposals for funding of nanotechnology-related projects by NIST's ATP program. ATP awarded approximately $8 million to industry for joint projects with NIST related to nanotechnology in fiscal year 98.

Focused Programs and Initiatives

Nanotechnology-based commercialization is expected to increase dramatically over the next decade. This will require NIST to respond with new standards and measurement capabilities in support of U.S. trade. The nanotechnology initiative will allow NIST to exploit these nanotechnology approaches in many more modes of activity. The research will lead to the development of a suite of intrinsic standards based on the manipulation and measurement at or near single atoms, electrons, and molecules. This new set of measurement and standards tools will enable NIST to anticipate industry requirements and respond at the same rapid pace as the technological innovations.

Research Opportunities

The main categories of research opportunity for DOC and NIST are to develop the measurement and standards infrastructure to support U.S. industry development and commercialization of nanotechnology, and to perform economic and foreign assessment studies. Major themes in fiscal year 2001 include nanodevices and biotechnology for quantum level measurement and calibration; magnetic measurements and standards research; nanoscale characterization: measurement systems, approaches and algorithms; standard data and materials; and nanoscale manipulation for synthesis and fabrication of measurement systems and standards.

2. DOD: Key Areas of DOD Interest in Nanostructures and Nanodevices

Main Themes

The Department of Defense depends on advanced technology to maintain U.S. military superiority. Rapid, complete, and accurate information is a key advantage in warfare. To meet this need, DOD has been the primary Federal supporter of electronics and materials sciences. The NSF Federal Funds for R&D Report (Meeks 1997) states that the DOD accounted for 64% of Federal technology base investment in electrical engineering in 1997, 36% in metallurgy and materials, and 55% in all engineering. Nanotechnology provides an opportunity for continued rapid progress in these fields.

Modern electronics and electro-optics are revolutionizing information acquisition, processing, storage, and display. The benefits to DOD of ongoing nanoscience R&D will include network centric warfare (near instantaneous collection and dissemination of information worldwide), augmented use of uninhabited combat vehicles (only surveillance platforms were available in 1999) to limit human casualties and enhance performance, and training aides (e.g., virtual reality tools to simulate real war without placing troops and equipment in actual physical danger).

<u>Information acquisition/processing/storage/display</u>

- Faster processors (smaller size enables higher speed)
- Denser memories (terabit per cm^2 implies 10 nm pixels)
- Lower-power, more affordable electronics (more innovative, more functional devices on chip)
- Sensor suites on a chip (NEMS, MEMS/electronics processing)
- Optoelectronic capabilities (giga-terahertz, multispectral, photonics)

Modern warfare is practiced with sophisticated weapons and platforms whose performance requirements far exceed normal commercial practices; higher-performance, more affordable materials are continually required to maintain military advantage.

Materials performance and affordability

- Extended life & maintenance (failure mechanisms initiate at nanometer scales)
- New properties in nanostructured materials (quantum and interface effects)
- Biomimetics (biological materials are frequently nanostructured composites)

Information, weapons, and platforms are tools; human performance ultimately determines warfare results. Improvements in monitors/controls of human physiology for performance assessment, casualty care, and susceptibility to chemical/biological agents are all important.

Bioengineering

- Chemical-bio-warfare defense (secondary/tertiary supramolecular structure key to sensing/protection)
- Casualty care (miniaturized devices to sense/actuate, protein manipulation)
- Personnel condition monitor/stimulus

Focused Programs and Initiatives

DOD has identified nanoscience as one of its six science and technology strategic research areas. The DOD S&T program is deliberately decentralized to more effectively couple the research to Service needs. Each Service—Air Force, Army, Navy/Marine Corps—has its own basic research program, funded by a Service budget line. These Service programs are largely single-investigator projects at universities; some additional work is performed at the Service laboratories. In addition, DARPA has a program to accelerate the development of promising but high-risk technologies, and DOD has a program to support research benefiting all the services. The DARPA and DOD programs are more likely to be multi-person, targeted at specific technology goals, and have well-defined lifetimes (3-5 years). Present DARPA and DOD programs targeting nanotechnology include the following:

- DARPA: Ultra Electronics, '91-'99; Advanced Microelectronics, '97-'00; Molecular Level Printing, '98-'00; Virtual Integrated Prototyping, '97-'00; Molecular Electronics, '99-'00; Terahertz (QD), '99-'02; Advanced Lithography, indefinite.
- MURI (Multidisciplinary University Research Centers): Mesoscale Patterning for Smart Material Systems, ARMY, Princeton U., '94-'99; Materials and Processing at the Nanometer Scale, AF, USC, '95-'00; Nanoscale Devices and Novel Engineered Materials, AF, U. Florida, '95-'00; Cluster Engineered Materials, ARMY, Northwestern U., '97-'02; Photonic Bandgap Engineering, ARMY, UCLA, '96-'01; Quantum Structures for Thermoelectric Applications, NAVY, UCLA, '97-'02; S&T of Nanotube-based Materials and Devices, NAVY, UNC, '98-'03; Self-Assembled Semiconductors: Size/Distribution Control, AF, USC, '98-'03; Computational Tools to Design/Optimize Nanodevices, AF, U. Minnesota, '98-'03; Engineering of Nanostructures and Devices, ARMY, Princeton U., '98-'03; Low-power High Performance Nanoelectronic Circuits, NAVY, Arizona State U., '98-'03; Science Base for Nanolithography, ARMY, U. New Mexico, '99-'04; Hybrid Molecular and Spin Semiconductors, ARMY, Purdue U., '99-'04.

Research Opportunities

Nanoelectronics and nanoelectromechanical systems (NEMS) will require a significant investment from DOD. A science base must be established to provide new approaches to affordable fabrication of quality nanostructures, understanding of the physical/chemical properties of those structures, the physics of new device concepts, and system architectures to exploit those new concepts in circuits. High-performance materials based on nanostructures require investment in new concepts for nanocluster formation, high surface area materials such as aerogels, measurement of interfacial properties, and physics/chemistry of materials failure mechanisms (materials failure initiates at the nanoscale). The considerable non-DOD investments in medicine and health will inevitably lead to enhanced casualty care and human performance; however DOD must pay attention to adapting these advances to its specific needs, especially for chemical/biological warfare defense.

3. DOE: Nanotechnology Research in the Department of Energy

Main Themes

The Department of Energy has two interests in nanometer-scale research: first, understanding the science and engineering of nanoscale systems, and second, development and use of nanotechnology-based systems in DOE technology programs such as defense programs and waste management. Nanotechnology covers the broad base of almost all scientific fields and fits between classical mechanics and quantum mechanics. Nanoparticles include viruses, genes, particles, structures, neurons, chemicals, solid state crystals, small magnets, and more. Our uncertainty regarding the science of these systems is large. There are large gaps in the knowledge of nanoscale physics, engineering, and physical properties, such as heat transfer, electron motion, collective phenomena, assembly of systems, controlling systems, impurity effects, and diagnostics. New scientific research on these systems will open many doors to new capabilities in DOE-related areas. In materials science, many new opportunities will be opened that will have profound ramifications for energy and the economy in such areas as electronics, solar energy, smart materials, ultracapacitors, magnets for motors and generators, batteries, and microfilters for chemical processing, water treatment, and waste entrainment.

Focused Programs and Initiatives

Already the Basic Energy Science (BES) materials science program has initiated basic research in several areas of nanoscale materials, including theory, computer simulations, structure and properties, materials processing, and new materials. Research is being carried out on a wide variety of materials, including nanotubes, fullerenes, semiconductors, nanocrystals, magnet precipitates, quantum dots and nanoclusters, magnet particles, and organically templated self-assembled structures.

In the BES engineering program, which is much smaller than the material science program, work has already begun on diagnostics to measure small nanoparticles in air, on surfaces, and in water. This will be very important in controlling environmental contamination in a society that will use nanoparticles in large amounts. The engineering program is also active in studying electron and heat transfer and physical properties of nanosystems and the fabrication of systems, including quantum dots and nanoelectronics. In engineering, nanotechnology will have major impact on diagnostics of systems, robotics, remote controls, environmental protection and knowledge, computer development, mass storage, smaller cheaper systems, systems failure, security, and other areas.

Among the new areas of opportunity in chemistry that are being explored in BES are catalysis, where small particles have long been the focus of researchers, separation science, molecular-level detection, and photochemical energy conversion. Among the opportunities that nanoscale chemistry offers is the possibility that this size scale will enable an understanding of the relationship between colligative (bulk) properties and molecular properties. In the technology programs at DOE, nanotechnology plays an increasing role.

4. DOT: Enhanced Awareness and Application of Nanotechnologies in Transportation

Main Themes

Transportation offers many opportunities for application of nanotechnology products and processes, such as nanostructured coatings, sensors for infrastructure, and smart materials. Within the Department of Transportation (DOT), the focus of research and development is on advancing applications of innovative enabling technologies, including nanotechnology. This focus is necessarily shaped by the department's mission, which is primarily operational, regulatory, and investment-driven.

The research conducted by DOT's operating administrations (the United States Coast Guard, Federal Aviation Administration, Federal Highway Administration, Federal Railroad Administration, Federal Transit Administration, Maritime Administration, National Highway Traffic Safety Administration, and Research and Special Programs Administration) is at the applied or developmental end of the spectrum and typically addresses relatively short-term, specific problems and opportunities. As specific nanotechnology research results are achieved relevant to DOT needs and responsibilities, the department will seek to encourage and incorporate them into its focused research and technology activities.

Focused Programs and Initiatives

DOT is actively involved in outreach efforts to ensure that the largely private and state-level transportation sector is aware of the potential value of current technological advances and is thereby motivated to exploit these opportunities. DOT chairs the NSTC Technology Committee (under which the IWGN falls) and its Transportation R&D Subcommittee. These committees are responsible for facilitating coordination of relevant research activities across the Federal Government and for fostering mutual awareness between the technology and transportation communities. As nanoscience and technology and related processes evolve, DOT will encourage and promote their application throughout the nation's transportation sector.

Research Opportunities

Relevant application areas and research opportunities for nanotechnology in transportation include atomic-level coating of metallic surfaces to achieve super-hardening, low friction, and enhanced corrosion protection; "tailored" materials with reduced life-cycle cost, greater strength-to-weight, and longer service life; and nanotechnology-based energy storage and fuel systems. Other opportunities include smart materials that monitor and assess their own status and health and that of the system or subsystem of which they are a part; developments here will result in a wide range of sensors (nanoscale to microscale) suitable for incorporation into microprocessor-controlled subsystems and components.

5. NASA: Towards Advanced Miniaturization and Functionality

Main Themes

The international space station program and Mars exploration studies have defined technology and research needs that are critical to their respective individual goals. These include research on micro- and nanotechnologies, manufacturing processes, and advanced materials. The goals of all are focused on enabling humans to live and work in space or on a planet, enhancing performance, reducing cost, and maintaining the health and well-being of the crew. Working in pursuit of these goals, a team of 11 people from the NASA Ames Research

Center recently was presented the Feynman Prize for Molecular Nanotechnology by the Foresight Institute. Also in pursuit of these goals, NASA and the National Science Foundation have held cooperative discussions on nanotechnology, as well as on biomedical technology and bioengineering. The NASA Human Exploration and Development of Space enterprise requirements include the following categories:

- Manufacturing technology: Improve rapid prototyping using the stereolithography and fusion deposition modeling techniques to produce functional prototypes and working models, including use of single-wall nanotubes. Improve composites manufacturing using fiber placement, filament winding, lamination, and resin transfer molding techniques. Improve the manufacture and precision of miniature mechanical components and electronic assemblies, and develop micro- and nanotechnologies to manufacture components.
- Nanotechnology: Potential applications and manufacturing techniques of single walled carbon nanotubes (SWNT), or "buckytubes." Applications include structural reinforcement; electronic, magnetic, lubricating, and optical devices; and chemical sensors and biosensors.

NASA's other near-term thrusts involve technological outreach activities to the microelectronics industry. NASA is using established tools to develop autonomous devices that articulate, sense, communicate, and function as a network, extending human presence beyond the normal senses. These tools will allow humans to function more fully while monitoring and maintaining bodily health. These technologies can be applied to health, ecology, warfare, and recreation.

With applications of existing technologies, progress can be accomplished in all these areas, but only to a limited degree. In all fields, present and near-term devices pale in comparison to examples in nature. The efficiency of energy conversion in mitochondria or in photosynthesis has not been equaled by human inventions. The complexity and efficiency of a simple reptilian nervous system far outstrips modern computer-driven robots. Biological sensors such as a dog's nose or a bat's ears far outstrip any man-made sensing device.

And the brains that process the information have no equal in our laboratories.

Research Opportunities

In the coming decades we will investigate the design of nature's finest inventions. Building devices that are closer to these biological models will require using techniques and materials that would be unfamiliar to any modern fabricator. These techniques are likely to grow out of combined efforts of molecular biologists, material scientists, electrochemists and polymer chemists. This synergy will surely result in new specializations that have not yet been conceived. Compared to the single crystalline-based technologies we now rely upon, future nanodevices will greatly outperform our current technologies due to their small size, varied composition, and molecular precision. Successful application of these technologies will augment our natural abilities, and, while conforming to human limitations, expand the human experience throughout our solar system and into the reaches of interstellar space.

6. NIH: Challenges in Nanobiotechnology

Main Themes

Currently, the National Institutes of Health support individual research efforts in the area of nanotechnology. NIH-supported nanotechnology-related projects span a range of research areas. Examples include the following:

- Design of DNA lattices that readily assemble themselves into predictable, two-dimensional patterns. These arrays are composed of rigid DNA tiles, about 60 nm^2, formed by antiparallel strands of DNA linked together by a double-crossover motif analogous to the crossovers that occur in meiosis. The precise pattern and periodicity of the tiles can be modified by altering DNA sequence, allowing the formation of specific lattices with programmable structures and features at a nanometer scale. This research has the potential to lead to the use of designed DNA crystals as scaffolds for the crystallization of macromolecules, as materials for use as catalysts, as molecular sieves, or as scaffolds for the assembly of molecular electronic components or biochips in DNA-based computers.

- Development of methodologies to improve the resolution and reduce the time and cost of nucleic acid sequencing.
- Work on understanding the principles of self-assembly, at different dimensional levels, and component material interfaces that define the formation of tooth enamel and bone.
- Development of nanotubules, nanoparticles, and nanospheres as drug delivery system scaffolds and for colorimetric detection of polynucleotides.
- Development of biosensors for the detection of single and multilayered molecular assemblies.
- Studies on motor proteins that convert chemical energy to mechanical energy and vice versa.
- Use of nanoscale structures for the development of nanostructured matrices and nanocomposite materials.
- Development of technologies such as atomic force microscopy to better visualize and understand cellular nanostructures, such as protein-DNA assemblies and cell organelles.

Future Planned Activities Relevant to Nanotechnology at NIH

Nanotechnology projects are currently supported under numerous programs; however, NIH has not yet planned specific initiatives focused exclusively in this area.

7. NSF Perspective: Fundamental Research and Education for Nanotechnology

Main Themes

- <u>Scientific and engineering frontiers at nanoscale:</u> Develop fundamental knowledge of basic nanoscale phenomena and processes by promoting discovery in key areas at the interfaces between physics, chemistry, biology, and engineering. Examples include research on novel phenomena, synthesis, processing, and assembly at nanoscale; modeling and simulation at nanoscale;

creation of materials by design; functional engineering at nanoscale; exploratory research on biosystems at the nanoscale.

- Create a balanced infrastructure for nanoscale science, engineering, technology, and human resources in this field, including university-industry partnerships.
- Education: Nanotechnology provides new opportunities to promote education at the interfaces between physics, mathematics, chemistry, biology, and engineering.

Programs and Initiatives

NSF has had activities in the following areas:

- Nanoscale science and engineering research is supported within disciplinary programs in the NSF directorates for Mathematical and Physical Sciences, Engineering, Biology, Computer and Information Systems and Engineering, and other areas.
- The Advanced Materials and Processing Program included aspects of nanostructures, molecular self-assembly, and nanochemistry (1988-94).
- Ultrafine Particle Engineering: synthesis and processing of nanoparticles at high rates (1991-1998).
- The "National Nanofabrication Users Network" has had a focus on electronics, extending a MEMS top-down approach in the first four years; currently it is expanding in the entire nanotechnology field (1994-; four-year renewal in 1998).
- The network "Distributed Center for Advanced Electronics Simulations" (DesCArtES), with the main center in the University of Illinois, has a main focus on nanoelectronics modeling and simulation.
- NSF funds several other centers: Science and Technology Centers at the University of California Santa Barbara (QUEST, 1989-) and Cornell University (Nanobiotechnology, 1999-); MRSECs include University of Wisconsin–Madison, Johns Hopkins University, University of Kentucky; ERCs include the University of Illinois; and Industry-University Cooperative Research Centers include the Particle Center at Penn State.

- Nano-instrumentation, NANO-95, was focused on increasing the success rate in the acquisition and development of nano-instrumentation (1995).
- Education opportunities are funded in centers, in collaborations with industry, and in groups of young U.S. researchers working in Japan and Europe. Research Experience for Undergraduate sites include Stanford University, University of South Carolina, Cornell University, UCSB, and Penn State University. Course development initiatives are also supported.
- The initiative "Partnership in Nanotechnology: Functional Nanostructures" (NSF 98-20) funded small groups working on functional nanostructures; the initiative involved participation of other agencies, national laboratories, and industry.
- An STTR (small business technology transfer) Solicitation on Nanotechnology was issued in July 1998; awards were made in fiscal year 1999.
- University-industry partnerships are encouraged (using the models provided by the program, Grand Opportunities for Academic Liaison with Industry, or GOALI).
- Nanoscale Science and Technology was an area highlighted in the NSF budget for 1998/99. NSF expenditure was $75 million in fiscal year 1998, and $90 million in fiscal year 1999 (or ~ 3% of the NSF research budget).
- The initiative "Exploratory Research on Biosystems at the Nanoscale" (NSF 99-109) focuses on high-risk, high-payoff research on nanoscale processes in biological and bio-inspired systems.

Research Opportunities

Several topics with increased potential are investigation of new phenomena, properties and processes; interface with bio-synthesis approaches, bio-nanostructures, and biomedicine; theory and simulation techniques for predicting synthesis and behavior of clusters; multifunctional/adaptive nanostructures; time-effective, economical, high-rate production methods; and broad enabling tools with impact on other disciplines and technologies.

Initiatives are being considered in the following areas:

- Nano-biotechnology: biosystems, bio-mimetics and composites
- Interfaces in environment at nanoscale: small length-scale/long time-scale processes; functional interfaces between bio/inorganic, inorganic, and biostructures
- New paradigms of operation, synthesis, and fabrication: nanostructures "by design;" quantum realm; and exploratory computational principles—quantum, DNA, etc.
- Integration of systems and architectures at the nanoscale: integration at nanoscale and with other scales; multiscale and multiphenomenal modeling and simulations

12.7.2 Current Funding for Nanotechnology Research

Contact person: M.C. Roco, IWGN

The estimations of Federal Government support for nanotechnology research in fiscal years 1997 and 1999 are as shown in Table 12.1 (not including programs at DOT, USDA, EPA, the Treasury Department, or BMDO):

Table 12.1. Estimations of Federal Government Support for Nanotechnology

	Fiscal Year 1997 **(Siegel et al. 1998)**	**Fiscal Year 1999** **(IWGN survey in September 1999)**
NSF	$65M	$85M
DOD (including DARPA, ARO, AFOSR, ONR)	$32M	$70M
DOE	$7M	$58M
DOC (including NIST with ATP)	$4M	$16M
NASA	$3M	$5M
NIH	$5M	$21M
TOTAL	$116M	$255M

12.7.3 Priorities for Funding Agencies: A Point of View

Contact: G. Whitesides, Harvard University

Role of Federal Agencies

NSF. NSF has established its role as an important source of support for small, peer-reviewed single-investigator university programs. It also supports the materials research science and engineering centers (MRSECs); these organizations, or structurally related new ones, would provide a cost-effective way of supporting shared facilities and encouraging small collaborative programs in nanoscience.

DARPA. DARPA has historically been a major source of support for nanoscience. This agency has been particularly good at building focused scientific/technical communities (especially those that mix university, industry, and national labs), and at focusing enough support on specific areas to give those areas strong boosts. It would be the natural lead agency in an effort in nanoscience and nanotechnology, and it has established programs in the area.

Other DOD agencies. ONR has historically been a vital source of support for chemistry, physics, and materials science. It also has effective programs for building communities, although the programs are typically more university-centered than those in DARPA. ONR's budget would have to be supplemented to build a program in nanoscience that is large enough to have a significant impact on the field.

NIH. NIH is the obvious source of support for the development of nanostructures relevant to biomedical devices (for example, genomic and proteomic chips). It has historically been receptive only to so-called hypothesis driven research and has not been very active in supporting device fabrication, engineering, or fundamental science leading toward devices; however, this prejudice against engineering now seems to be disappearing. NIH would also be the plausible partner in shared support intended to integrate biologists/physiologists with device fabricators.

DOE. DOE has been an important source of support for specific classes of nanomaterials with implications for energy production (e.g., supported catalysts and zeolites) and for other areas of materials research (e.g., buckytubes). These activities would map well onto support for one of the important areas of nanoscience. There will be some requirement in nanoscience and technology for use of large facilities (for example, X-ray light sources for X-ray lithography and X-ray crystallography); supporting and managing these facilities within the Federal budget is usually the task of DOE.

DOC. DOC, through NIST, would be a plausible source of technical support for start-up and venture-backed companies in nanotechnology.

Startup/Venture Enterprises

One reason that biology and biotechnology research and development programs have been so effective in the United States has been that the difficult step of transferring technology from the universities has been accomplished, in part, by a specialized group of small, venture-backed, startup companies. In principle, the same structure could work to facilitate transfer of university science into commercial technology. It is not clear what the Federal Government can do to accelerate the formation of a venture capital community active in nanotechnology. Probably key issues are to develop intellectual property policies that make it attractive to transfer technologies into startups, and perhaps to provide (through DOD) protected markets for products just as they enter the marketplace.

Global Research

Nanotechnology—as an extension of microtechnology—is being explored by national-scale efforts in several countries. Among the leaders in such efforts are Japan and Israel; countries such as Finland are also emerging as powerful centers of microtechnology R&D, and these countries could be either leaders or fast followers into new technology areas. The United States had the luxury of developing microtechnology without serious competition until the early 1980s. We will not enjoy the same uncluttered environment in the development of nanotechnology; rather, the area will be fiercely competitive.

Priorities

- Build a broad program of R&D in nanoscience that includes research universities, relevant industry, and some of the national laboratories. Building a "community" focused on nanoscience/technology, and providing stable support for this community at a level high enough to allow the participating groups to reach a critical mass, are important objectives of public policy.

- Develop mechanisms for bridging communities interested in nanoscience that span the gaps between the physical and biological sciences.
- Develop policies explicitly designed to attract large companies as participants in programs of Federally funded groups. Without large company participation, technology development programs in nanoelectronics will probably fail at the stage of research planning and product definition.
- Develop a strategy for informal coordination of R&D among participating Federal agencies, centered at NSF.
- Maintain an active series of reports to Congress and the Office of Management and Budget, both to aid in educating policymakers about the progress, opportunities, and failures of the field, and to provide a general education about nanoscience at senior levels in the Government.
- Provide funds to push science that seems to offer the potential of developing into profitable technology rapidly to the point of manufacturable prototypes using focused, DARPA-style programs.
- Address the problems of public perception of threat from nanoscience by Federal Government regulatory programs and active programs in public education.

12.7.4 Functional Nanostructures: An Initiative and its Outcomes

Contact person: M.C. Roco, NSF

The NSF-wide initiative "Partnership in Nanotechnology on Functional Nanostructures" (NSF 98-20) was funded in fiscal year 1998. The initiative was addressed to interdisciplinary small-group projects, and only two proposals were accepted from each university. The NSF competition received 178 proposals and made 24 awards for a total of $13 million, of which 25% were GOALI projects, 37% had a formal international component, and 37% had direct contributions from other funding agencies. Programs in fourteen NSF divisions jointly funded the awards. The response has encouraged collaborative activities among NSF programs and other agencies.

12.8 REFERENCES

Kalil, T. 1999. Nanotechnology: Time for a national initiative? In *IWGN Workshop Proceedings*, January 27-29 (personal communication at the workshop).

Meeks, R.L. 1997. *Federal funds for research and development: Fiscal years 1995, 1996, and 1997.* Vol. 45, Detailed statistical tables. Arlington, VA: National Science Foundation, Division of Science Resources Studies. NSF 97-327 (http://www.nsf.gov/sbe/srs/nsf97327/start.htm).

Moore, D. 1999. Research priorities. Presentation at IWGN workshop, January 27-29 (personal communication at the workshop).

Roco, M.C. 1999. Research programs on nanotechnology in the world. Chapter 8 in *Nanostructure science and technology,* NSTC 1999 *(see below).*

Siegel, R.W., E. Hu, and M.C. Roco, eds. 1998. *R&D status and trends in nanoparticles, nanostructured materials, and nanodevices in the United States*. Baltimore: International Technical Research Institute, Loyola College, World Technology (WTEC) Division, Web site: http://itri.loyola.edu/nano/US.Review/. Also available as NTIS #PB98-117914.

Siegel, R.W., E. Hu, and M.C. Roco, eds. 1999. NSTC (National Science and Technology Council) Report. *Nanostructure science and technology.* Baltimore: International Technology Research Institute, World Technology (WTEC) Division. Web site: http://itri.loyola.edu/nano/IWGN.Worldwide.Study/. Also published by Kluwer Academic Publishers (1999).

APPENDICES

A. LIST OF PARTICIPANTS AND CONTRIBUTORS

IWGN

Casassa, Michael P.
Director, Program Office
NIST
Admin. Bldg., Rm. A1000
Gaithersburg, MD 20899

Genther-Yoshida, Phyllis
Associate Director, ITPTA
U.S. Dept. of Commerce
Room 4411 HCHB
14th and Constitution Ave., NW
Washington, DC 20230

Henkart, Maryanna P.
Directorate for Biology
National Science Foundation
4201 Wilson Blvd., Suite 655S
Arlington, VA 22230

John, Richard R.
Director, DOT Volpe National
Transportation Systems Center
55 Broadway, Kendall Square
Cambridge, MA 02142

Kraback, Tim
Jet Propulsion Laboratory
Building 179, Rm. 224A
4800 Oak Grove Rd.
M/S 179-224
Pasadena, CA 91109-8099

Kalil, Thomas A.
Special Special Assistant to the President
White House Economic Council
OEOB, Washington, DC 20502

Kirkpatrick, Kelly
Senior Policy Analyst, Office of Science
and Technology Policy
OEOB, Washington, DC 20502

Kousvelari, Eleni
Director, Biomimetics and Tissue
Engineering Program
National Institute of Dental and
Craniofacial Research
National Institutes of Health
Natcher Bldg. (Bldg. 45), Rm. 4AN-18A
Bethesda, MD 20892-6402

Kovatch, George
Chief, Advanced Vehicle Technologies
Division
DOT Volpe National Transportation
System Center;
55 Broadway, Kendall Square
Cambridge, MA 02142

Lacombe, Annalynn
DOT Volpe National Transportation
System Center;
55 Broadway, Kendall Square
Cambridge, MA 02142

Meyyappan, Meyya
NASA Ames Research Center
Mail Stop 229-3, Bldg 229, Room 214
Moffett Field, CA 94035

Mucklow, Glenn H.
National Aeronautics and Space
Administration
Office of Space Science
Headquarters Building
Washington DC 20546

Murday, James S.
Superintendent
Chemistry Division
Naval Research Laboratory
Code 6100
Washington, D.C. 20375-5342

Murphy, Edward
U.S. Dept. of Treasury

Pomrenke, Gernot S.
Mathematics and Space Sciences Directorate
Air Force Office of Scientific Research
Ballston Common Towers II
801 N. Randolph Street, Rm. 732
Arlington, VA 22203

Porter, Joan
NSTC, EOP, OSTP
Old Executive Office Bldg., Rm. 435
Washington, D.C. 20502

Price, Robert
U.S. Department of Energy
ER-15, Office of Basic Energy Sciences
1000 Independence Ave., S.W.
Code SC-15, Rm. E-438A
Washington, DC 20585

Radzanowski, Dave
Office of Management and Budget, EOP
New Executive Office Building
125 17th Street, N.W.
Washington, DC 20503

Roco, M.C.
Program Director
National Science Foundation
Division of Chemical and Transport Systems
4201 Wilson Blvd., Suite 525
Arlington, VA 22230

Schloss, Jeffery A.
Program Director
Technology Development Coordination
National Human Genome Research Institute
National Institutes of Health
Bldg 38A, Room 614
38 Library Drive
Bethesda, MD 20892-6050

Shull, Robert D.
NIST
Magnetic Materials
Bldg. 223, Rm. B152
Gaithersburg, MD 20899

Thomas, Iran L.
Division of Materials Sciences, Office of Basic Energy Sciences
SC-13, Rm. F-402
U.S. Department of Energy
19901 Germantown Road
Germantown, MD 20874-1290

Trew, Robert J.
Director of Research
Office of the Secretary of Defense Research and Engineering
4015 Wilson Blvd., Suite 209
Arlington, VA 22203

Venneri, Samuel
Chief Technologist
National Aeronautics and Space Administration
Headquarters Building, Rm. 9S13
Washington DC 20546-0001

Volintine, Brian G.
U.S. Department of Energy
Office of Industrial Technologies
EE-20, Rm. 5F-064
1000 Independence Ave., S.W.
Washington, D.C. 20585-0121

Weber, Thomas A.
Directorate for Mathematical and Physical Sciences
National Science Foundation
4201 Wilson Blvd., Suite 1065
Arlington, VA 22230

Other Govt. & NL contributors

Barhen, Jacob
Center for Engineering Systems Advanced Research (CESAR)
ORNL
PO Box 2008
Oak Ridge TN 37831-6355

Berson, Alan
NHLBI, NIH
Two Rockledge Centre, Room 9178
6701 Rockledge Drive MSC 7940
Bethesda, MD 20892

Brecher, Dr. Aviva
Transportation Strategic Planning and Analysis Office (DTS-24)
DOT/RSPA Volpe National Transportation Systems Center
55 Broadway
Cambridge
MA 02142-1093

Celotta, Robert J.
NIST Fellow and Group Leader, Electron Physics Group
NIST
B206 Metrology Building
100 Bureau Drive, MS 8412
Gaithersburg, MD 20899-8412

Dahl, Carol
Assistant to the Director
National Cancer Institute
Strategic Technologies Office
Bldg 31, Rm. 11AO3
31 Center Drive, MSC 2590
Bethesda, MD 20892-2590

Dixon, David A.
Pacific Northwest Laboratory
P.O. Box 999
MS K1-83
Richland WA 99352

Eastman, Jeffrey A.
Argonne National Lab.
Materials Science Div.
9700 S. Cass Avenue
Bldg. 212
Argonne, IL 60439

Edwards, James L.
Deputy Assistant Director
Directorate for Biological Sciences
National Science Foundation
4201 Wilson Blvd.
Arlington D.C. 22230

Everitt, H.
U.S. Army Research Office
Physics Division
U.S. Army Research Office
Box 12211, RTP, NC 27709-2211

Exarhos, Gregory
Manager, Material Research Section
Batelle Pacific Northwest Laboratory
P.O.Box 999,MS K2-44
Richland, WA 99352

Gilman, Jeffrey W.
Fire Science Division
Polymers Building, Room A265
NIST
100 Bureau Drive, Stop 8652
Gaithersburg, MD 20899-8652

Ginley, David
National Renewable Energy Laboratory
1617 Cole Boulevard
Golden, CO 80401-3393

Hagnauer, Gary
Army Research Laboratory
Weapons & Materials Research Directorate
Attn: AMSRL-WM-M (Materials)
Aberdeen Proving Ground, MD 21005

Hasslacher, Brosl
Los Alamos National Laboratory
Institute for Physical Sciences
713 Meadow Lane
Los Alamos, NM 87544

Heetderks, William J.
NINDS, NIH
FED/802
National Institutes of Health
Bethesda, MD 20892

Hinkley, Jeffrey A.
NASA Langley Research Center
Materials Division
Mail Stop 226
6a West Taylor Street
Hampton, VA 23681-2199

Kelley, Chris
NHLBI/NIH
31 Center Drive MSC 2490
Bethesda, MD 20892

Kelley, Richard D.
U.S. Department of Energy
Materials Sciences Division
SC-13, Rm. F-421
Building: GTN
Washington, DC 20545

Kung, Harriet
Los Alamos National Laboratory
Materials Science and Technology
Division
Mail Stop G755
Los Alamos, NM 87545

Li, DeQuan
Los Alamos National Laboratory
Bioscience Division
Mail Stop G755
Los Alamos, NM 87545

Lowe, Terry C.
Los Alamos National Laboratory
Materials Science Laboratory, TA3-1698
Mail Stop G754
Los Alamos, NM 87545

Michalske, Terry A.
Sandia National Laboratories
Surface and Interface Science
Department
P.O. Box 5800
Albuquerque, NM 87185-1413

Nelson, David L.
National Science Foundation
Division of Materials Research
4201 Wilson Blvd., Rm. 1065N
Arlington, VA 22230

Newbury, Dale
NIST
Bldg. 222, Rm A113
Gaithersburg, MD 20899

Picraux, S. Tom
Director, Physical & Chemical Sciences
Center
Sandia National Labs
P.O. Box 5800
Albuquerque, NM 87185-1427

Poehlein, Gary
Division Director, CTS/ENG
National Science Foundation
4201 Wilson Blvd.
Arlington, VA 22230

Ramsey, Mike
Oak Ridge National Laboratory
P.O. Box 2008
1 Bethel Valley Rd.
Oak Ridge, TN 37831-6142

Samara, George A.
Sandia National Laboratories
Nanostructures and Advanced Materials
Chemistry Department, MS-1421
P.O. Box 5800
Albuquerque, NM 87185-1421

Seyfert, Vicki L.
NIAID, NIH
Building SB, Room 4A21
31 Center Drive
Bethesda, MD 20892-2425

Stroscio, Michael A.
Chief Scientist
Office of the Director
U.S. Army Research Office
4300 S. Miami Blvd.
Research Triangle Park, NC 27709-2211

Tassey, Gregory
Senior Economist
National Institute of Standards and
Technology
100 Bureau Drive, Stop 1060
Gaithersburg, MD 20899-1060

Tilden, Mark W.
Los Alamos National Laboratory
Physics Division
Mail Stop D454
Los Alamos, NM 87545

Toomerian, Benny
Jet Propulsion Laboratory
CISM C Center for Integrated
Microsystems
4800 Oak Grove Drive, MIS 303-310
Pasadena, CA 91109

Ward, Keith B.
ONR, Code 341
800 N. Quincy Street
Arlington, VA 22217-5660

Warren, William
DARPA/DSO
3701 N. Fairfax Dr.
Arlington, VA 22203-1714

Watson, John
NHLBI, NIH
Building 31, Room 5A49
31 Center Drive MSC 2490
Bethesda, MD 20892

Wong, Eugene
Assistant Director for Engineering
National Science Foundation
4201 Wilson Blvd.
Arlington D.C. 22230

Academic contributors

Aksay, Ilhan
Princeton University
A313 Chemical Engineering Quadrangle
Princeton, NJ 08544-5263

Alivisatos, Paul
University of California at Berkeley
Chemistry Dept.
B62 Hildebrand 6911
Berkeley, CA 94720-6911

Andres, Ronald P.
School of Chemical Engineering
Purdue University
West Lafayette, IN 47907-1283

Ballantyne, Joseph M.
Director, Cornell Nanofabrication Facility
Cornell University
M105 Knight Lab
Ithaca, NY 14853

Bawendi, Moungi G.
Chemistry Dept.
MIT
Rm. 6-223
77 Massachusetts Ave.
Cambridge, MA 02139

Block, Stephen M.
Stanford University
Department of Biological Sciences
Gilbert Building, Room 109
371 Serra Mall
Stanford, CA 94305-5020

Brown, Patrick
Stanford University
Biochemistry Department
Beckman B400, Mail Code 5307
Stanford, CA, 94305-5307

Chelikowsky, James R.
University of Minnesota
Chemical Engineering & Materials Science Dept.
Room 151 Amundson Hall, Mail Stop 0531
421 Washington Ave SE
Minneapolis, MN 55455

Chou, Stephen
Princeton University
Dept. of Electrical Engineering
B412 Engineering Quad
Princeton, NJ 08544
609-258-4416
chou@ee.princeton.edu

Chu, Steve
Stanford University
Applied Physics Department
Varian Bldg Room 230
Stanford, CA. 94305-4085

Colbert, Daniel T.
Rice University
Dept. of Chemistry
Smalley Research MS100
241 Butcher Hall
P.O. Box 1892
Houston, TX 77251-1892

Craighead, Harold
Professor of Applied and Engineering Physics
Cornell University
211 Clark Hall
Ithaca, NY 14853

Cummings, Peter T.
Dept. of Chemical Engineering
University of Tennessee
419 Dougherty Engineering
Knoxville, TN 37996

Dresselhaus, Mildred S.
MIT Institute Professor
Professor of Physics and Electrical Engineering
MIT
Rm. 13-3005
77 Mass Ave.
Cambridge, MA 02139

Ellis, Arthur B.
Univ. of Wisconsin-Madison
Dept. of Chemistry
750 University Ave.
Madison, WI 53706-1490

Ferrari, Mauro
Ohio State University
Professor of Engineering Science
270 Bevis Hall
1080 Carmack Road
Columbus OH 43210-1002

Flagan, Richard
CALTECH
M/C 210-41
Pasadena, CA 91125

Friedlander, Sheldon K.
Chemical Engineering Department
UCLA, 405 Hilgarg Avenue
Los Angeles, CA 90024-1592

Goddard, William A., III
California Institute of Technology
Chemistry Dept.
M/C 139-74
Pasadena, CA 91125

Guntherodt, Hans-Joachim
Institute of Physics
University of Basel
Klingelbergstr. 82
CH-4056 Basel
Switzerland

Heath, James R.
UCLA
Department of Chemistry & Biochemistry
607 Circle Drive South
Box 951569
Los Angeles, CA 90095-1569

Hess, Karl
Beckman Institute
University of Illinois - Urbana-Champaign
405 N. Mathews Ave
Urbana, IL 61801

Hu, Evelyn
Center for Quantized Electronic Structures
Science & Technology Center
University of California
Santa Barbara, CA 93106

Jelinski, Lynn
Vice Chancellor for Research and Graduate Studies
Louisiana State University
Office of Research and Graduate Studies
240 Thomas Boyd Hall
Baton Rouge, LA 70803

Kear, Bernard
Rutgers Univ.
Ceramics and Materials Engineering
Rutgers University,
607 Taylor Road,
Piscataway, NJ 08854

Kiehl, Richard A.
University of Minnesota
Electrical and Computer Engineering Dept.
EE/CSci Building, Room 6-129
200 Union St. SE
Minneapolis, MN 55455-0154

Klabunde, Kenneth J.
Kansas State University
Department of Chemistry
Willard Hall
Manhattan, KS 66506

Koehler, Karl A.
Purdue Research Foundation
Division of Sponsored Program Development
HOVDE 1021, Rm. 313
West Lafayette, IN 47907-1021

Lavernia, Enrique J.
UC Irvine
Department of Chemical and Biochemical Engineering and Materials Science
Irvine, CA 92697

Leong, Kam
Johns Hopkins University
Biomedical Engineering
720 Rutland Avenue
Ross Bldg. 726
Baltimore, MD 21205

Leuenberger, Hans
Vice President, Swiss Academy of Engineering Sciences
University of Basel
Department of Pharmacie
CH-4051 Basel, Switzerland

Majumdar, Arun
Associate Professor
University of California, Berkeley
Mechanical Engineering
6169 Etcheverry
Berkeley, CA 94720-1740

Merz, James
Vice President for Graduate Studies and Research
University of Notre Dame
208 Hurley Hall
Notre Dame, IN 46556-5641

Najafi, Khalil
Univ. of Michigan
EECS Dept.
2402 EECS Bldg.
1301 Beal Ave.
Ann Arbor, MI 48109-2122

Navrotsky, Alexandra
Univ. of California at Davis
Dept. of Chemical Engineering and Materials Science
4440 Chemistry Annex
Davis, CA 95616-5294

Oberdorster, Gunter
Director of the Division of Respiratory Biology and Toxicology
University of Rochester
Department of Environmental Medicine
575 Elmwood Ave.
Rochester, NY. 14642

Odde, David J.
University of Minnesota
Department of Biomedical Engineering
7-104 Basic Sciences/Biomedical Engineering Building
312 Church St., S.E.
Minneapolis, MN 55455

Petroff, Pierre
Department Of Materials
University of California at Santa Barbara
Eng III Building
Santa Barbara, CA 93106

Pollack, Lois
Cornell University
Physics Department
Clark Hall 168
Ithaca, NY 14853

Porod, Wolfgang
Notre Dame University
Electrical Engineering Dept.
268 Fitzpatrick Engineering
Notre Dame, IN 46556

Pui, David
University of Minnesota
Mechanical Engineering Department
111 Church St. S.E.
Minneapolis, MN 55455

Quate, C.F.
Stanford University
Ginzton Lab
Stanford, CA 94305-4085

Raber, Douglas J.
Director, Board of Chemical Sciences and Technology
National Research Council, (NAS-273),
2101 Constitution Avenue N.W.
Washington, D.C. 20418

Ratner, Mark A.
Northwestern University
Chemistry Dept.
2145 Sheridan Rd.
Evanston, IL 60208-3113

Reed, Mark A.
Chair, Electrical Engineering Dept.
Yale University
P.O. Box 208284
New Haven, CT 06520-8284

Requicha, Aristides A.G.
Laboratory for Molecular Robotics
University of Southern California
941 West 37th Place
Los Angeles, CA 90089-0781

Ruoff, Rod
Dept of Physics
Washington University
CB 1105, One Brookings Drive
St. Louis, MO 63130-4899

Schlick, Tamar
New York University
Dept. of Chemistry
251 Mercer Street
New York, NY 10012

Siegel, Richard W.
Head, Materials Science and Engineering Department
Rensselaer Polytechnic Institute
110 Eighth Street
Troy, New York 12180-3590

Smith, Temple F.
Boston University
Biomedical Engineering Dept.
Biomolecular Engineering Research Center
36 Cummington Street
Boston, MA 02215

Stormer, Horst
Columbia University
Physics Dept.
2960 Broadway
704 Pupin, M/C 5206
New York, NY 10027

Stucky, Galen D.
Department of Chemistry
University of California, Santa Barbara
Santa Barbara, CA 93106-9510

Stupp, Samuel I.
Departments of Materials Science and Chemistry
Northwestern University
2225 N. Campus Drive
Evanston, IL 60208

Sussman, Michael
University of Wisconsin
Biotechnology Center
425 Henry Mall
Madison, WI 53706-1580

Tirrell, Matthew V.
Richard A. Auhll Professor and Dean
College of Engineering
University of California, Santa Barbara
Santa Barbara, CA 93106

Tour, James M.
Rice University
Center for Nanoscale Science and Technology
MS 222, P.O. Box 1892
Houston, TX 77251-1892

Van Der Weide, Daniel W.
Dept. of Electrical and Computer Engineering
University of Wisconsin-Madison
1415 Engineering Drive
Madison, WI 53706-1691

Whitesides, George
Harvard University
Department of Chemistry
12 Oxford St.
Cambridge, MA 02138

Ying, Jackie
Dept. of Chemical Engineering, Rm. 66-544
Massachusetts Institute of Technology
Cambridge, MA 02139-4307

Private sector contributors

Avouris, Phaedon
IBM Thomas J. Watson Research Center
P.O. Box 218
Yorktown Heights, NY 10598

Bauer, David R.
Sr. Staff Technical Specialist
Ford Motor Company
Research Laboratory, MD3182
P.O. Box 2053
Dearborn, MI 48121

Blair, John
DOB
JBX Technologies, Inc.
25 Moore Road
Wayland, MA 01778

Brannon-Peppas, Lisa
President, Biogel Technology, Inc.
P.O. Box 681513
9521 Valparaiso Court
Indianapolis, IN 46268

Cader, Tahir
Energy International, Inc.
127 Bellevue Way SE, Suite 200
Bellevue, Washington 98004-6229

Canton, James
President, Institute for Global Futures
2084 Union Street
San Francisco, CA 94123

Cavin, Ralph K. III
Vice President, Research Operations
Semiconductor Research Corp.
P.O. Box 12053
Research Triangle Park, NC 27709-2053

Chapman, Kevin
Senior Director, Combinatorial Chemistry
Merck Research Labs
Mail Code RY 123-230
P.O. Box 2000
Rahway, NJ 07065

Cody, George D.
Exxon Corporate Research
Route 22 East Clinton Township
Annandale, New Jersey 08801

Cohen, Roger
Director of Strategic Planning and Programs
Exxon Corporate Center
Exxon R&E Company
Annandale, NJ 08801

Cox, Donald M.
Exxon Research & Engineering Co. (retired)
14 Beechwood Place
Watchung, NJ 07060

Doering, Robert R.
TI Senior Fellow
Silicon Technology Development
Texas Instruments Inc.
P.O. Box 650311
MS 3730
Dallas, TX 75265

Drew, Stephen W.
Merck & Co.
P.O. Box 2000
Rahway, NJ 07065

Eigler, Donald
IBM Almaden Research Center
650 Harry Rd., Mail Stop D1
San Jose, CA 95120-6099

Ellenbogen, James
Mitre Corporation
1820 Dolly Madison Ave. M/S Room W635
McLean, VA 22102

Frazier, Gary
Raytheon Systems Company
Mail Station 35
13532 N. Central Expressway
Dallas, Texas 75243

Garces, Juan M.
Research Scientist
Dow Corporate R&D
Building 1776
Midland, MI 48674

Gardner, Colin
Vice President
Merck Research Laboratories
P.O. Box 4
WP 78-202
West Point, PA 19486

Goronkin, Herb
Motorola EL 508
2100 East Elliott Road
Tempe, AZ 85284

Grochowski, Edward
IBM Almaden Research Center
650 Harry Road
San Jose CA 95120

Heller, Michael J.
Chief Technical Officer
Nanogen
10398 Pacific Center Court
San Diego, CA 92121

Helms, Jeffrey
Ford Motor Company
20000 Rotunda, MD-3182
Dearborn, MI 48121-2053

Helvajian, Henry
Sr. Scientist
Aerospace Corporation
P.O. Box 92957
Los Angeles, CA 90009-2957

Jasinski, Joseph M.
IBM Research
T. J. Watson Research Center
P. O. Box 218
Yorktown Heights NY 10598

Kresge, Charles T.
Paulsboro Technical Center
Mobil Technology Company
P.O. Box 480
Paulsboro, NJ 08066

Kukonnen, Carl
ViaSpace Technologies
2400 Lincoln Ave.
Altadina, CA 91001

Lee, Stephen Craig
Senior Research Investigator
Monsanto Company
700 Chesterfield Village Parkway AA4C
St. Louis, MO 63196

Loutfy, Raouf O.
President
MER Corporation
7960 S. Kolb Rd.
Tucson, AZ 85706

Manian, Bala S.
Biometric Imaging
1025 Terra Bella Ave
Mountain View, CA 94043

Mendel, John
Eastman Kodak
1669 Lake Avenue
Bldg. 35, 2nd Floor, Room 257, Mail Code 23701
Rochester, N.Y. 14652-3701

Newberry, Deb
Executive Director, Space Systems
General Dynamics Information Systems
8800 Queen Ave. S. BLCS1P
Bloomington, MN 55431

Peercy, Paul
President
Semi/Sematech
2796 Montopolis Drive
Austin, TX 78741-6499

Popper, Steven W.
Senior Economist
Assoc. Director
S&T Policy Institute
RAND Corporation

Rawal, Suraj
Lockheed Martin Astronautics
PO Box 179
M/S 3085 DC
Denver, CO 80201

Salzman, David
Polychip, Inc.
4812 Auburn Ave.
Bethesda, MD 20814

Strulovici, Berta
Merck and Co.
Dept. of Automated Biotechnology
NW-1
502 Louise Lane
North Wales, PA. 19454

West , Paul
ThermoMicroscopes
1171 Borregas Ave.
Sunnyvale, CA 94089

Williams, R. Stanley
Principal Laboratory Scientist
Hewlett Packard Laboratories
3500 Deer Creek Rd.
MS 26U-12
Palo Alto, CA 94304-1126

Wiltzius, Pierre
Bell Labs
Lucent Technologies
700 Mountain Avenue
Rm. ID428
Murray Hill, NJ 07974

Wise, John J.
Mobil (retired)
4343 Province Line Road
Princeton, NJ 08540

Ziolo, Ronald F.
Xerox Corporation
Corporate Research & Technology
0114-39D
800 Phillips Road
Webster, NY 14580

B. IWGN REFERENCE MATERIALS

This is a list of key publications and conference proceedings sponsored by IWGN and participating agencies in the area of nanoscience, engineering and technology in the last two years.

The following documents have been prepared by the NSTC/CT/IWGN:

1. *National Nanotechnology Initiative,* internal NSTC/CT/IWGN report, reviewed by the President's Committee of Advisors on Science and Technology (PCAST) Nanotechnology Panel. Expected release in Feb. 2000 (see http://www.nsf.gov/nano).

2. *Nanostructure Science and Technology* (NSTC Report), Siegel et al. 1999, eds., worldwide study on status and trends; available on the Web: http://itri.loyola.edu/nano/IWGN.Worldwide.Study/, on CD-ROM from WTEC, and hard cover publication by Kluwer Academic Publishers (1999).

3. *Nanotechnology Research Directions: IWGN Workshop Report* (NSTC report), Roco et al. 1999, eds., providing input from the academic, private sector and government communities; available on the Web: http://itri.loyola.edu/nano/IWGN.Research.Directions/ (this report).

4. *Nanotechnology – Shaping the Word Atom by Atom* (NSTC report), I. Amato, brochure for the public (available on the Web at: http://itri.loyola.edu/nano/IWGN.Public.Brochure/).

Additional information on the National Nanotechnology Initiative will be posted on the Web at: http://www.nsf.gov/nano.

The following publications/proceedings prepared by different agencies since 1997 address specific scientific topics, technological issues or areas of relevance in nanoscale science and engineering:

- *R&D Status and Trends in Nanoparticles, Nanostructured Materials, and Nanodevices in the United States* (includes review of U.S. funding), sponsored by 7 agencies in 1997, Proceedings published in January 1998, R. Siegel, E. Hu and M.C. Roco, eds., WTEC, on the Web: http://itri.loyola.edu/nano/US.Review/

- *NSF-NIST Conference on Nanoparticles* (NSF and NIST, 1997), Proceedings, D.T. Shaw, M.C. Roco and R. Shull, eds.

- *Selfassembling* (NSF, 1997), Proceedings (1998), M. Tirrell, Ed.

- *Interdisciplinary Macromolecular Science and Engineering* (NSF, 1997), Workshop report (Brochure in colors, 1998), S. Stupp, ed.

- *Molecular Modeling and Simulation* (NSF, 1997), Workshop Report 1998, P.T. Cummings, ed., Website: http://flory.engr.utk.edu/nsf/

- *US-EC Workshop on Nano-biotechnology* (NSF, 1997), L. Jelinski, Workshop Report (Brochure in colors, 1998), Website: http://www.bio.cornell.edu/nanobiotech/nbt.htm/

- *A Research Needs Assessment: Future Use of Nanofabricated Materials in Energy Applications* (DOE, 1998), Y.M. Chiang and J.Y. Ying.

- *Thermal-chemical simulations* (DOE, 1998), Proceedings, D. Dixon and P.T. Cummings, eds.

- *BECON Proceedings* (NIH, 1998), Workshop report, H.G. Craighead et al., eds., Website: http://www.nih.gov/grants/becon/report_19980228.pdf, page 39.

- *Nanotubes and Nanoelectronics* (SRC/NASA Ames, 1998), Workshop Proceedings, D. Herr, M. Meyyappan and V. Zhirnov, eds.

- *Nanotechnology for the Soldier System* (ARO/DOD, 1998), Proceedings, T. Tassinnari, ed.

- *Ultra Electronics Program Review* (DARPA/DOD, 1998), Proceedings, Website: http://web-ext2.darpa.mil/eto/ULTRA/index.html, G. Pomrenke, ed.

- *NANOSpace 98*, NASA Workshops on Nano-Micro Technology, 1998, Houston, Proceedings on CD-ROM in 1999.

- *Nanoscale Science, Engineering and Technology Research Directions*, Oak Ridge National Laboratory, 1999 (with focus on research opportunities and challenges for national laboratories), D.H. Lowndes et al., eds.

- *Condensed-Matter and Materials Physics* (National Research Council, 1999), publication supported by the DOC and NSF, National Academy Press, Washington, D.C., 1999.

C. GLOSSARY

A/D	Analog (to) digital
AES	Auger electron spectroscopy
AFM	Atomic force microscopy
AFOSR	Air Force Office of Scientific Research
AIM	Atomic imaging and manipulation
ALE	Atomic layer epitaxy
ATM	Asynchronous transmission mode
ATP	(U.S., Dept. of Commerce) Advanced Technology Program
ATP	Adenosine triphosphate (supplies energy for biochemical cellular processes)
bcc	body-centered cubic
BEEM	Ballistic electron emission microscope/microscopy
BEP	(Treasury Department) Bureau of Engraving and Printing
BES	(Department of Energy) Basic Energy Science Directorate
CAD-T	Computer-aided design and test
CAT (scan)	Computerized axial tomography
CD	Compact disk
Cermet	Ceramic-metal composite
CFU	Colony forming units
CMOS	Complementary metal-oxide semiconductor
CMP	Chemical mechanical polishing
CNT	Carbon nanotube
CPSE	Collaborative problem-solving environments
CPU	Central processing unit (e.g., of a computer)
CRT	Cathode ray tube
CVD	Chemical vapor deposition
D/A	Digital (to) analog
DARPA	Department of Defense Advanced Research Projects Agency
DesCArtES	(NSF funded) Distributed Center for Advanced Electronics Simulations
DFT	Density functional theory
DRAM	Dynamic random access memory
DURIP	(DOD program) Defense University Research Instrumentation Program
EELS	Electron energy loss spectroscopy
ERC	(NSF) Engineering and Research Center
ESM	Electronic support measures
ESPRIT	European Commission's information technologies program
EUV	Extreme ultraviolet

EXAFS	Extended X-ray absorption fine structure
fcc	Face centered cubic
FDA	Force discrimination assay
FED	(Japan) R&D Association for Future Electron Devices
FEL	Free electron laser
FET	Field effect transistor
FRET	Fluorescence resonance energy transfer
FWHM	Full width, half maximum
GMR	Giant magnetoresistance
GOALI	(NSF) Grant Opportunities for Academic Liaison with Industry
HDS	Hydrodesulfurization
HFET	Heterojunction field-effect transistor
HPMA	N-2-hydroxypropyl methacrylamide
HREELS (41)	High- resolution electron energy loss spectroscopy (EELS)
HREM	High-resolution electron microscopy
HVOF	High velocity oxygen fuel
IP	Ionization potential
IPO	Initial public offering
IR	Infrared
IWGN	(U.S., NSTC) Interagency Working Group on Nanoscience, Engineering and Technology
JPL	(U.S., NASA) Jet Propulsion Laboratory (Pasadena)
JSC	(NASA) Johnson Space Center
LANL	Los Alamos National Laboratory
LARC	(NASA) Langley Research Center
LED	Light-emitting diode
LFM	Lateral force microscopy
LISA	Lithographically-induced self-assembly
MBE	Molecular beam epitaxy
MD	Molecular dynamics
MEL-ARI	(Europe, ESPRIT) Advanced Research Initiative in Microelectronics
MEMS	Microelectromechanical systems
MFM	Magnetic force microscope/scopy
MITI	(Japan) Ministry of International Trade and Industry
MOMBE	Organometallic molecular beam epitaxy
MRI	Magnetic resonance imaging
MRM	Magnetic resonance microscope/scopy
MRSEC	(NSF) Materials Research Science and Engineering Center(s) (network)
MS	Mass spectrometry

M-TAS	Micro(scale)-total analysis system
MURI	(DOD) Multidisciplinary University Research Initiative
nano-TAS	Nano(scale) total analysis system
NASA	National Aeronautics and Space Administration
NCSA	(U.S.) National Computational Science Alliance (for high-performance computing)
NCSL	Nanocrystal super lattices
NDR	Negative differential resistance
NEMS	Nanoelectromechanical systems
NIL	Nanoimprint lithography
NIST	National Institute of Standards and Technology
nm	Nanometer
nM	Nanomanipulator
NMR	Nuclear magnetic resonance
NNUN	(U.S.) National Nanofabrication Users Network
NP	Non-deterministic polynomial
NQR	Nuclear quadrupole resonance
NREL	National Renewable Energy Laboratory
NSOM	Near-field scanning optical microscope/microscopy
NST	(DOD) Nanometer Science and Technology
NSTC	(U.S.) National Science and Technology Council
OMB	(U.S.) Office of Management and Budget
OMVPE	Organometallic vapor phase epitaxy
OPS	Operations per second
OPTO	(Europe/ESPRIT/MEL-ARI) Optoelectronic Interconnects for Integrated Circuits
PARC	(U.S./Xerox) Palo Alto Research Center
PDMS	Polydimethylsiloxane
PEEM	Photoemission electron microscope/microscopy
PI	Principal investigator
PMMA	Polymethylmethylmethacrylate
PVR	Peak-to-valley ratio
PVS	Physical vapor synthesis
QCA	Quantum cellular automata
QMOS	Quantum metal oxide semiconductor
RAM	Random access memory
RCT	(JPL) Revolutionary Computing Technologies Program
RTD	Resonant tunneling diode
SAM	Self-assembled monolayer
SAMMS	Self-assembled monolayers on mesoporous supports
SBIR	(U.S.) Small Business Innovative Research (program)

SCM	Scanning capacitance microscope/scopy
SECM	Scanning electrochemical microscope/scopy
SEED	Self-electro-optic effect device
SEM	Scanning electron microscopes/microscopy
SET	Single electron transistor
SFA	Surface force apparatus
SIA	(U.S.) Semiconductor Industry Association
SMT	(U.S., Rutgers Univ.) Strategic Materials Program
SOAC	System on a chip
SPICE	Simulation Program with Integrated Circuit Emphasis (computer-aided design tool)
SPM	Scanning probe microscope/scopy
SRAM	Static random access memory
STC	(NSF) Science and Technology Center
STEM	Scanning transmission electron microscope
SThM	Scanning thermal microscope/scopy
STM	Scanning tunneling microscope/microscopy
STS	Scanning tunneling spectroscopy
STTR	(U.S.) Small Business Technology Transfer (program)
SWNT	Single-wall nanotube (also, "buckeytube")
TAS	Total analysis systems ("chem lab on a chip")
TEM	Transmission electron microscope/microscopy
Teramak	R&D program at Hewlett-Packard
TSRAM	Tunneling-based static RAM
UPS (41)	Ultraviolet photoelectron spectroscopy
URI	(U.S., DOD) University Research Initiative
UV	Ultraviolet
UV/VIS (41)	Ultraviolet/visible light
VCSEL	Vertical cavity surface emitting laser
XMDC	X-ray magnetic circular dicroism
XPS	X-ray photoelectron spectroscopy

D. INDEX OF AUTHORS

E. INDEX OF MAIN TOPICS

Nanotechnology Research Directions

Preface to the Reprinted Version

The first edition of this book was intended to provide strategic guidance for the United States R&D community as a new technology develops out of scientific and engineering advances at the nanometer scale. It was utilized as the basis for the U.S. National Nanotechnology Initiative (NNI), and it has influenced R&D strategy discussions in many other countries as well. Nanoscience and engineering was defined beyond just the study of small things: new properties and functions of matter emerge at the nanoscale, and thus nanotechnology will require that matter be intentionally crafted to control composition, size, shape and position at length scales from the nanometer to macroscopic dimensions. The book presented a snapshot of the various disciplines influenced by nanoscale science at the beginning of 1999. The rate of discovery in these disciplines has accelerated dramatically since the reports were compiled, and potential applications are now better defined. The establishment of national research and educational priorities for nanoscience and nanoengineering in virtually all industrialized countries in the past few years is a sign of the broad recognition of the potential scientific, technological and societal benefits of this emerging field. Nanotechnology is joining molecular biology and information technology as the key technologies for the 21st Century.

This report is the synthesis of opinions from about 150 leading experts almost equally distributed among academic, industrial, and government scientists, including those from research laboratories and funding agencies. It identified challenges and opportunities in nanoscale science and engineering research and development, and addressed key issues in order to develop a wide range of commercial applications. The report makes specific recommendations on how to develop a balanced infrastructure for nanotechnology R&D, advance critical research areas, and nurture the scientific and technical workforce for the future. These recommendations stood as the basis of the National Nanotechnology Initiative announced by President Clinton in January 2000, and approved by U.S. Congress in November 2000.

The book was written in a modular fashion to allow readers with any background to learn something useful about Nanotechnology and its scientific, educational, commercial and societal implications. It has twelve chapters, covering fundamental scientific issues, investigative tools, synthesis and processing of nanostructures, key areas for commercial utilization, research infrastructure needs, and strategies for funding agencies. These topics have relevance to researchers and educators, as well as to industrial and government policy makers. The basic concepts and illustrations maintain their relevance since the publication of the first edition.

M.C. Roco, R.S. Williams and P. Alivisatos

THE WHITE HOUSE

WASHINGTON

September 27, 1999

Dear Colleague:

In August 1999, the National Science and Technology Council's (NSTC) Interagency Working Group on Nanoscience, Engineering, and Technology (IWGN) released its first report, entitled *Nanostructure Science and Technology*. That document provided a basis for the Federal government to assess how to make strategic research and development (R&D) investments in this emerging field of nanotechnology through the formulation of national R&D priorities and a strategy for state, local, and Federal government support.

This IWGN Workshop Report, *Nanotechnology Research Directions,* builds upon the foundation provided in the first report and incorporates a vision for how the nanotechnology community -- Federal agencies, industries, universities, and professional societies -- can more effectively coordinate efforts to develop a wide range of revolutionary commercial applications. It incorporates perspectives developed at a January 1999 IWGN-sponsored workshop by experts from universities, industry, and the Federal government. This report identifies challenges and opportunities in the nanotechnology field and outlines the necessary steps on how advances made in nano-science, engineering, and technology can help to boost our nation's economy, ensure better healthcare, and enhance national security in the coming decade.

Preparing for the challenges of the new millennium requires strategic investments. *Nanotechnology Research Directions* will help our nation develop a balanced R&D nanotechnology infrastructure, advance critical research areas, and nurture the scientific and technical workforce of the next century.

Sincerely,

Neal Lane

Neal Lane
Assistant to the President
for Science and Technology